# MIXTURE MODELS

## STATISTICS: Textbooks and Monographs



**Vol. 1:** The Generalized Jackknife Statistic, *H. L. Gray and W. R. Schucany*
**Vol. 2:** Multivariate Analysis, *Anant M. Kshirsagar*
**Vol. 3:** Statistics and Society, *Walter T. Federer*
**Vol. 4:** Multivariate Analysis: A Selected and Abstracted Bibliography, 1957-1972, *Kocherlakota Subrahmaniam and Kathleen Subrahmaniam* (out of print)
**Vol. 5:** Design of Experiments: A Realistic Approach, *Virgil L. Anderson and Robert A. McLean*
**Vol. 6:** Statistical and Mathematical Aspects of Pollution Problems, *John W. Pratt*
**Vol. 7:** Introduction to Probability and Statistics (in two parts), Part I: Probability; Part II: Statistics, *Narayan C. Giri*
**Vol. 8:** Statistical Theory of the Analysis of Experimental Designs, *J. Ogawa*
**Vol. 9:** Statistical Techniques in Simulation (in two parts), *Jack P. C. Kleijnen*
**Vol. 10:** Data Quality Control and Editing, *Joseph I. Naus* (out of print)
**Vol. 11:** Cost of Living Index Numbers: Practice, Precision, and Theory, *Kali S. Banerjee*
**Vol. 12:** Weighing Designs: For Chemistry, Medicine, Economics, Operations Research, Statistics, *Kali S. Banerjee*
**Vol. 13:** The Search for Oil: Some Statistical Methods and Techniques, *edited by D. B. Owen*
**Vol. 14:** Sample Size Choice: Charts for Experiments with Linear Models, *Robert E. Odeh and Martin Fox*
**Vol. 15:** Statistical Methods for Engineers and Scientists, *Robert M. Bethea, Benjamin S. Duran, and Thomas L. Boullion*
**Vol. 16:** Statistical Quality Control Methods, *Irving W. Burr*
**Vol. 17:** On the History of Statistics and Probability, *edited by D. B. Owen*
**Vol. 18:** Econometrics, *Peter Schmidt*
**Vol. 19:** Sufficient Statistics: Selected Contributions, *Vasant S. Huzurbazar (edited by Anant M. Kshirsagar)*
**Vol. 20:** Handbook of Statistical Distributions, *Jagdish K. Patel, C. H. Kapadia, and D. B. Owen*
**Vol. 21:** Case Studies in Sample Design, *A. C. Rosander*
**Vol. 22:** Pocket Book of Statistical Tables, *compiled by R. E. Odeh, D. B. Owen, Z. W. Birnbaum, and L. Fisher*
**Vol. 23:** The Information in Contingency Tables, *D. V. Gokhale and Solomon Kullback*

**Vol. 24:** Statistical Analysis of Reliability and Life-Testing Models: Theory and Methods, *Lee J. Bain*
**Vol. 25:** Elementary Statistical Quality Control, *Irving W. Burr*
**Vol. 26:** An Introduction to Probability and Statistics Using BASIC, *Richard A. Groeneveld*
**Vol. 27:** Basic Applied Statistics, *B. L. Raktoe and J. J. Hubert*
**Vol. 28:** A Primer in Probability, *Kathleen Subrahmaniam*
**Vol. 29:** Random Processes: A First Look, *R. Syski*
**Vol. 30:** Regression Methods: A Tool for Data Analysis, *Rudolf J. Freund and Paul D. Minton*
**Vol. 31:** Randomization Tests, *Eugene S. Edgington*
**Vol. 32:** Tables for Normal Tolerance Limits, Sampling Plans, and Screening, *Robert E. Odeh and D. B. Owen*
**Vol. 33:** Statistical Computing, *William J. Kennedy, Jr. and James E. Gentle*
**Vol. 34:** Regression Analysis and Its Application: A Data-Oriented Approach, *Richard F. Gunst and Robert L. Mason*
**Vol. 35:** Scientific Strategies to Save Your Life, *I. D. J. Bross*
**Vol. 36:** Statistics in the Pharmaceutical Industry, *edited by C. Ralph Buncher and Jia-Yeong Tsay*
**Vol. 37:** Sampling from a Finite Population, *J. Hajek*
**Vol. 38:** Statistical Modeling Techniques, *S. S. Shapiro*
**Vol. 39:** Statistical Theory and Inference in Research, *T. A. Bancroft and C.-P. Han*
**Vol. 40:** Handbook of the Normal Distribution, *Jagdish K. Patel and Campbell B. Read*
**Vol. 41:** Recent Advances in Regression Methods, *Hrishikesh D. Vinod and Aman Ullah*
**Vol. 42:** Acceptance Sampling in Quality Control, *Edward G. Schilling*
**Vol. 43:** The Randomized Clinical Trial and Therapeutic Decisions, *edited by Niels Tygstrup, John M. Lachin, and Erik Juhl*
**Vol. 44:** Regression Analysis of Survival Data in Cancer Chemotherapy, *Walter H. Carter, Jr., Galen L. Wampler, and Donald M. Stablein*
**Vol. 45:** A Course in Linear Models, *Anant M. Kshirsagar*
**Vol. 46:** Clinical Trials: Issues and Approaches, *edited by Stanley H. Shapiro and Thomas H. Louis*
**Vol. 47:** Statistical Analysis of DNA Sequence Data, *edited by B. S. Weir*
**Vol. 48:** Nonlinear Regression Modeling: A Unified Practical Approach, *David A. Ratkowsky*
**Vol. 49:** Attribute Sampling Plans, Tables of Tests and Confidence Limits for Proportions, *Robert E. Odeh and D. B. Owen*
**Vol. 50:** Experimental Design, Statistical Models, and Genetic Statistics, *edited by Klaus Hinkelmann*
**Vol. 51:** Statistical Methods for Cancer Studies, *edited by Richard G. Cornell*
**Vol. 52:** Practical Statistical Sampling for Auditors, *Arthur J. Wilburn*
**Vol. 53:** Statistical Signal Processing, *edited by Edward J. Wegman and James G. Smith*
**Vol. 54:** Self-Organizing Methods in Modeling: GMDH Type Algorithms, *edited by Stanley J. Farlow*
**Vol. 55:** Applied Factorial and Fractional Designs, *Robert A. McLean and Virgil L. Anderson*
**Vol. 56:** Design of Experiments: Ranking and Selection, *edited by Thomas J. Santner and Ajit C. Tamhane*
**Vol. 57:** Statistical Methods for Engineers and Scientists. Second Edition, Revised and Expanded, *Robert M. Bethea, Benjamin S. Duran, and Thomas L. Boullion*
**Vol. 58:** Ensemble Modeling: Inference from Small-Scale Properties to Large-Scale Systems, *Alan E. Gelfand and Crayton C. Walker*
**Vol. 59:** Computer Modeling for Business and Industry, *Bruce L. Bowerman and Richard T. O'Connell*
**Vol. 60:** Bayesian Analysis of Linear Models, *Lyle D. Broemeling*

**Vol. 61:** Methodological Issues for Health Care Surveys, *Brenda Cox and Steven Cohen*

**Vol. 62:** Applied Regression Analysis and Experimental Design, *Richard J. Brook and Gregory C. Arnold*

**Vol. 63:** Statpal: A Statistical Package for Microcomputers – PC-DOS Version for the IBM PC and Compatibles, *Bruce J. Chalmer and David G. Whitmore*

**Vol. 64:** Statpal: A Statistical Package for Microcomputers – Apple Version for the II, II+, and IIe, *David G. Whitmore and Bruce J. Chalmer*

**Vol. 65:** Nonparametric Statistical Inference, Second Edition, Revised and Expanded, *Jean Dickinson Gibbons*

**Vol. 66:** Design and Analysis of Experiments, *Roger G. Petersen*

**Vol. 67:** Statistical Methods for Pharmaceutical Research Planning, *Sten W. Bergman and John C. Gittins*

**Vol. 68:** Goodness-of-Fit Techniques, *edited by Ralph B. D'Agostino and Michael A. Stephens*

**Vol. 69:** Statistical Methods in Discrimination Litigation, *edited by D.H. Kaye and Mikel Aickin*

**Vol. 70:** Truncated and Censored Samples from Normal Populations, *Helmut Schneider*

**Vol. 71:** Robust Inference, *M. L. Tiku, W. Y. Tan, and N. Balakrishnan*

**Vol. 72:** Statistical Image Processing and Graphics, *edited by Edward J. Wegman and Douglas J. DePriest*

**Vol. 73:** Assignment Methods in Combinatorial Data Analysis, *Lawrence J. Hubert*

**Vol. 74:** Econometrics and Structural Change, *Lyle D. Broemeling and Hiroki Tsurumi*

**Vol. 75:** Multivariate Interpretation of Clinical Laboratory Data, *Adelin Albert and Eugene K. Harris*

**Vol. 76:** Statistical Tools for Simulation Practitioners, *Jack P. C. Kleijnen*

**Vol. 77:** Randomization Tests, Second Edition, *Eugene S. Edgington*

**Vol. 78:** A Folio of Distributions: A Collection of Theoretical Quantile-Quantile Plots, *Edward B. Fowlkes*

**Vol. 79:** Applied Categorical Data Analysis, *Daniel H. Freeman, Jr.*

**Vol. 80:** Seemingly Unrelated Regression Equations Models : Estimation and Inference, *Virendra K. Srivastava and David E. A. Giles*

**Vol. 81:** Response Surfaces: Designs and Analyses, *Andre I. Khuri and John A. Cornell*

**Vol. 82:** Nonlinear Parameter Estimation: An Integrated System in BASIC, *John C. Nash and Mary Walker-Smith*

**Vol. 83:** Cancer Modeling, *edited by James R. Thompson and Barry W. Brown*

**Vol. 84:** Mixture Models: Inference and Applications to Clustering, *Geoffrey J. McLachlan and Kaye E. Basford*

**Vol. 85:** Randomized Response: Theory and Techniques, *Arijit Chaudhuri and Rahul Mukerjee*

**Vol. 86:** Lognormal Distributions: Theory and Applications, *edited by Edwin L. Crow and Kunio Shimizu*

**Vol. 87:** Handbook of Biopharmaceutical Statistics, *edited by Karl E. Peace*

## ADDITIONAL VOLUMES IN PREPARATION

# MIXTURE MODELS

## Inference and Applications to Clustering

Geoffrey J. McLachlan
Kaye E. Basford
University of Queensland
Brisbane, Queensland
Australia

MARCEL DEKKER, INC. New York and Basel

**Library of Congress Cataloging-in-Publication Data**

McLachlan, Geoffrey J.
Mixture models : inference and applications to clustering / Geoffrey J. McLachlan, Kaye E. Basford.
p. cm. — (Statistics, textbooks and monographs ; v. 84)
Bibliography: p.
Includes indexes.
ISBN 0-8247-7691-7
1. Mixture distributions (Probability theory) 2. Cluster analysis. I. Basford, Kaye E. II. Title. III. Series.
QA276.7.M39 1987
519.5'3—dc19
87-18931
CIP



MARCEL DEKKER, INC.
270 Madison Avenue, New York, New York 10016

Current printing (last digit):
10 9 8 7 6 5 4 3 2 1

PRINTED IN THE UNITED STATES OF AMERICA

*To*

*Beryl, Jonathan, and Robbie*

*Alan, Lee, and Jay*

# Preface

The purpose of this book is to highlight the important role of finite mixture distributions in modeling heterogeneous data, with the focus on applications in the field of cluster analysis. With the increasing interest on a model-based approach to clustering, the use of finite mixture models for this purpose has been the subject of close scrutiny recently. The practical applications of mixture models are presented against the background of the existing literature for inference undertaken in a finite mixture framework. Attention is concentrated on the fitting of mixture models by a likelihood-based approach, using maximum likelihood where appropriate. This approach would appear in general to be superior to other methods of fitting mixture models, and the EM algorithm provides a convenient way for the iterative computation of solutions of the likelihood equation. However, it has only been in relatively recent times with the wide availability of high-speed computers that serious consideration has been given to the fitting of mixture models to data of more than one dimension.

The emphasis here is on mixtures of normal distributions as these models are most widely employed in practice, as well as most frequently studied. In keeping with the applied spirit of the book, there is also an account of items such as the detection of atypical observations, the assessment of model fit, and robust estimation for finite mixture models.

More specifically, the use of mixture models is considered in two different contexts:

1. Where there is a genuine finite mixture structure.

2. Where there is no *a priori* knowledge of any formal group structure on the underlying population, but where one wishes to cluster the data into a number of groups. A convenient labeling scheme may be all that is sought, though usually it is hoped that the particular grouping obtained may shed some light on the phenomena of interest.

Against the background of case 1, a comprehensive account is provided of likelihood estimation for finite mixture models. Both theoretical and practical issues are discussed. In case 1, the sole aim may be to estimate the parameters of the component distributions and the proportions in which the components of the mixture occur. Indeed, the estimation of the mixing proportions is a problem that frequently arises in practice, and is treated here in some depth. In other instances, the estimation of the parameters of the mixture may be just a means for constructing a rule for assigning the data at hand to the externally existing components of the mixture. An example is in discriminant analysis, where there has been an impediment to the collection of adequate training data from each of the possible components.

The major theme of this work, however, is on case 2. With a mixture model, a clustering can be effected by assigning the data on the basis of the estimated posterior probabilities of component membership. One application of the mixture likelihood approach to clustering considered is its grouping of treatment means in an analysis of variance, where it provides a concise way of summarizing differences between the treatments. Also, it is demonstrated how the mixture approach can be used to cluster so-called three-way data, where one of the modes (classes) is to be partitioned into groups on the basis of the other two modes considered simultaneously. As the mixture likelihood approach to clustering is model-based, it allows estimation and hypothesis testing within the framework of standard statistical theory, although theoretical difficulties still remain with some problems. One such difficult problem addressed is assessing the smallest number of components in a mixture consistent with the data. This problem arises with the question of how many clusters there are or how many classes there are in a latent class model. Another problem considered is the assessment of the effectiveness of the clustering produced by the mixture likelihood approach for the given data. For both these problems, proposals are considered that make use of the powerful bootstrap technique.

Numerous examples involving the statistical analyses of real data sets are presented throughout to demonstrate the various applications of finite mixture models. These data sets are taken from the fields of agriculture, botany, education, medicine, and zoology. Also, examples are described from other fields, in particular, pattern recognition in association with the analysis of remotely sensed data from the Landsat earth satellite. A FORTRAN listing of computer programs is given in the Appendix for the fitting of normal mixture models to data sets from a variety of experimental designs.

This monograph should appeal to both applied and theoretical statisticians, as well as to investigators working in the many diverse fields in which relevant use can be made of finite mixture models. It is assumed that the reader has a fair mathematical or statistical background. Readers interested primarily in the applications of mixture models may wish to bypass the theoretical discussions of the more esoteric issues of likelihood estimation in a finite mixture framework.

The book can be used as a source of references on work of either a practical or theoretical nature on finite mixture models. An attempt has been made to cover the main results in the area, but a thorough review is not intended. Over 350 references are given, although of course not all of them are specific to mixtures. There are also selected references from other areas such as data, discriminant, and cluster analyses to complement the presentation.

Portions of the material in this book have been used by the first author to complement the standard texts in his lectures on multivariate analysis to fourth-year honors (graduate level) students. Besides its obvious relevance to the theory and applications of finite mixture models and to parametric approaches in cluster analysis, the book contains results which are of interest to multivariate data analysis in other contexts, but which are unable to be explored in any depth in texts giving a broad coverage of multivariate analysis. For example, the systematic treatment here of the assessment of multivariate normality and homoscedasticity for the component distributions of the mixture and for the detection of atypical observations covers not only the usual cluster analysis setting, but also the more straightforward case where the unclassified data are from a genuine mixture and where there are additional data of known origin from each of its components. The methods given, which proceed conditionally on the components of the mixture for this latter case, can also be used to assess the customary distributional assumptions for the multivariate analysis of variance and discriminant analysis.

The first author would like to acknowledge gratefully the financial support of the Australian Research Grants Scheme during part of the writing of the book. The same author also wishes to offer his appreciation to the British Science and Engineering Research Council for financial support to attend the workshop on Pattern Recognition and Statistics held in Edinburgh, July, 1985. As this trip coincided with the initial drafting of the book, discussions on the latter with participants at the workshop were most useful.

We wish to thank the authors and owners of copyright for permission to reproduce published tables and figures. Finally, we express our appreciation to the staff at Marcel Dekker, Inc. for their help in the production of this volume.

**Geoffrey J. McLachlan**
**Kaye E. Basford**

# Contents

# MIXTURE MODELS

# 1
# General Introduction

## 1.1 INTRODUCTION

Mixtures of distributions, in particular normal, have been used extensively as models in a wide variety of important practical situations where data can be viewed as arising from two or more populations mixed in varying proportions. Recent problems which have been addressed by normal mixture models include the identification of outliers (Aitkin and Tunnicliffe Wilson, 1980) and the investigation of the robustness of certain statistics to departures from normality, for example, the sample correlation coefficient as studied by Srivastava and Lee (1984). In addition to this latter role of assessing the performances of estimators in nonnormal situations, normal mixtures have been used of course in the development of robust estimators. For example, under the contaminated normal family as suggested by Tukey (1960), the density of an observation was taken to be a mixture of two univariate normal densities with the same means but where the second component had a greater variance than the first. This family was introduced to model a population which follows a normal distribution except on those few occasions where a grossly atypical observation is recorded. Huber (1964) subsequently considered more general forms of contamination of the normal distribution in the development of his robust M-estimators of a location parameter.

It is in the field of cluster analysis, however, that mixtures of distributions, normal or otherwise, are being increasingly used. Various examples of clustering on the basis of normal mixtures may be found in the articles by Basford and McLachlan (1985a to d). Mixtures of discrete distributions have been exploited as a clustering technique, for instance in latent class analysis. In the models proposed by Lazarsfeld and Henry (1968), the observed correlations between several dichotomous variables are explained by the superpopulation being a mixture of "latent classes" within each of which the variables are independently distributed. An example from Aitkin, Anderson and Hinde (1981) will be considered in Chapter 3.

In this book we shall be highlighting the use of mixture models fitted according to the likelihood approach as a way of providing an effective clustering of various data sets under a variety of experimental designs. Under the mixture likelihood approach to clustering, it is assumed that the observations on the entities to be clustered are from a mixture of an initially specified number of populations or groups in various proportions. By adopting some parametric form for the density function in each underlying group, a likelihood can be formed in terms of the mixture density, and unknown parameters estimated by consideration of the likelihood. A probabilistic clustering of the entities can then be obtained in terms of their estimated posterior probabilities of group membership. An outright assignment of the entities to the groups can be achieved by allocating each entity to the group to which it has the highest estimated posterior probability of belonging. Although the estimates of the parameters, and hence of the posterior probabilities, may not be reliable if the sample size is not large, a satisfactory clustering as given by this outright assignment may still be achieved (Ganesalingam and McLachlan, 1979a). This last comment is in the context where the data are from a genuine mixture, so that it is not inappropriate to speak about the true values of the parameters and posterior probabilities.

It can be seen that the mixture likelihood approach to clustering is model based in that the form of the density of an observation in each of the underlying populations has to be specified. Hawkins, Muller and ten Krooden (1982, page 353) commented that most writers on cluster analysis "lay more stress on algorithms and criteria in the belief that intuitively reasonable criteria should produce good results over a wide range of possible (and generally unstated) models." For example, the trace $\mathbf{W}$ criterion, where $\mathbf{W}$ is the pooled within cluster covariance matrix, is predicated on normal populations with spherical covariance matrices, but as they pointed out, many users would apply this criterion even in the face of evidence of nonspherical clusters, or equivalently, would use Euclidean distance as a metric. They strongly supported the increasing emphasis on a model based approach to clustering. Indeed, as remarked by Aitkin, Anderson and Hinde (1981) in the reply to the discussion of their paper, "when clustering samples from a population, no cluster method is *a priori* believable without a statistical model." Concerning the use of mixture models to represent nonhomogeneous populations, they noted in their paper that "Clustering methods based on such mixture models allow estimation and hypothesis testing within the framework of standard statistical theory."

Within this framework, two fundamental questions associated with the application of a given clustering technique can be assessed formally for the mixture likelihood approach. Firstly, on the question of the number of clusters, a test for the smallest number of groups compatible with the data can be formulated in terms of the likelihood ratio criterion, although unfortunately it does not have its usual asymptotic distribution under the null hypothesis. New attempts at resolving this difficult but important problem shall be discussed later. Secondly, on the question of the effectiveness of the clustering obtained for the specified number of groups, it is possible using the estimated posterior probabilities of group membership, to form estimates of the overall and individual correct allocation rates which would exist if the clustering were assumed, for this particular purpose, to represent an externally existing partition. Basford and McLachlan (1985a) proposed the use of these estimated rates after bias correction according to the bootstrap as a measure of the strength of the clustering. This facility for assessing the performance of the mixture approach to clustering is highly desirable, and will be examined here in some detail.

With a parametric formulation of finite mixture models, matters of initial concern include the estimation of the unknown parameters and the resulting fit of the mixture for the parametric forms adopted for its component densities. These and other issues as distinct from the subsequent clustering properties of finite mixture models will also be addressed in this book. Much attention is to be given to inference problems posed with likelihood estimation in a mixture framework. Practical applications of mixture models are to be considered too, where the estimates of the unknown parameters are of interest solely in their own right and not just as part of a preliminary statistical analysis leading to a clustering of the data at hand. For example, an important step in the automatic cytoscreening for cervical cancer is the estimation of the proportions of various types of cells present in a PAP smear specimen. Another example concerns crop acreage estimation, using remotely sensed data from mixtures of several crops. There are many other examples in the recent literature to provide further evidence of the widespread applicability of finite mixture distributions. They include the modeling of crime and justice data by Harris (1983) using mixtures of geometric and negative binomial distributions, the description of wind shear data by Kanji (1985), and the regression analysis of competing risks data by Larson and Dinse (1985).

## 1.2 HISTORY OF MIXTURE MODELS

Decomposing a finite mixture of distributions is a very difficult problem, as can be seen by looking at the solution based on the method of moments put forward by Karl Pearson (1894) in the case of a mixture of two univariate distributions with unequal variances. It requires the solution of a ninth degree polynomial equation, although Cohen (1967) subsequently showed how this could be circumvented through an iterative process which involves solving a cubic equation for a unique negative root. This approach was suggested by the solution in the case of equal variances where the estimates depend uniquely on the negative root of a cubic equation constructed from the first four moments; see Charlier and Wicksell (1924) and Rao (1952, Section 8b.6). However, Tan and Chang (1972) and Fryer and Robertson (1972), among others, have shown that the method of moments is inferior to likelihood estimation for this problem.

As speculated by Fowlkes (1979) in his study of diagnostic plotting procedures for the detection of univariate mixtures, it was probably because of the intractability of the moment estimators and the absence of modern computer technology that attention was focused on graphical techniques for mixtures during the early and mid-1900's. Pioneering work on these techniques was undertaken by Harding (1948) and Cassie (1954), and continued by Bhattacharya (1967) and Wilk and Gnanadesikan (1968) among others. Tarter and Silvers (1975) presented a graphical procedure based on the properties of the bivariate Gaussian density function, while Chhikara and Register (1979) developed a numerical classification technique based on computer aided methods for the display of the data.

With the advent of high speed computers, attention was turned to likelihood estimation of the parameters in a mixture distribution. The first use of this principle for a mixture model has been attributed to Rao (1948) who used Fisher's method of scoring for a mixture of two univariate distributions with equal variances. However, Butler (1986) recently pointed out that Newcomb (1886), predating even Pearson's (1894) early attempt on mixture models with the method of moments, suggested an iterative reweighting scheme which can be interpreted as an application of the EM algorithm of Dempster, Laird and Rubin (1977) to compute the maximum likelihood estimate of the common mean of a mixture in known proportions of a finite number of univariate normal populations with known variances. Also, Butler (1986) noted that Jeffreys (1932) used essentially the EM algorithm in iteratively computing the estimates of the means of two univariate normal populations which had known variances and which were mixed in known proportions. Following Rao's (1948) paper, likelihood estimation appears not to have been pursued further until Hasselblad (1966, 1969) addressed the problem, initially for a mixture of $g$ univariate normal distributions with equal variances, and then for mixtures of distributions from the exponential family. The former case was also considered briefly by Behboodian (1970) and its multivariate analogue by Wolfe (1967, 1970, 1971) and Day (1969) in major papers. Day (1969) concentrated on the solution for two normal populations with the same covariance matrix, while Wolfe (1967, 1970, 1971) dealt with an arbitrary number of normal heteroscedastic populations as well as mixtures of multivariate Bernoulli distributions for use in latent class analysis. As with Hasselblad (1966, 1969), their solutions were presented in an iterative form corresponding to particular applications of the EM algorithm of Dempster, Laird and Rubin (1977). However, it was not until the latter had formalized this iterative scheme in a general context through their EM algorithm, that the convergence properties of the likelihood solution for the mixture problem were established on a theoretical basis.

The likelihood approach to the fitting of mixture models, in particular normal mixtures, has since been utilized by several authors, including Dick and Bowden (1973), Hosmer (1973a and b, 1974, 1978), O'Neill (1978), Ganesalingam and McLachlan (1978, 1979a, 1980a), and Aitkin (1980a). The work in these and more recent references will be discussed shortly, commencing with the formal definition of this approach in Section 1.4.

It can be seen from the above outline that the finite mixture problem has quite a lengthy history. By now, there is an extensive literature, and so it not possible to provide an exhaustive bibliography here. However, we have attempted in this book to cover the main results on finite mixture models, in particular the more recent work, which is concerned mainly with the likelihood based fitting of these models. Earlier references, such as those associated with the initial development of mixture models through the method of moments, may be found in the comprehensive bibliographies in the books on finite mixture distributions by Everitt and Hand (1981) and Titterington, Smith and Makov (1985). Included in the latter is a table containing 90 references on direct applications of mixture models, giving in each case the field of the application, the form of the data, the component distributions adopted, and the estimation method used. In addition, there are the review articles by Gupta and Huang (1981), Holgersson and Jorner (1978), and Redner and Walker (1984), and the encyclopedia entries by Blischke (1978) and Everitt (1985). Also, McLachlan (1982a) reviewed the likelihood approach to the fitting of mixture models, concentrating on their role as a device for clustering.

Before proceeding with the formulation of finite mixture models, leading in particular to the formal definition of the mixture likelihood approach to clustering, a brief review of the general classification problem is given to place this approach in perspective.

## 1.3 BACKGROUND TO THE GENERAL CLASSIFICATION PROBLEM

Firstly, it is important to establish a standard terminology to describe the data sets which will be considered. Carroll and Arabie (1980) introduced "a taxonomy of measurement data," in which a mode is defined as a particular class of entities, and an $N$-way array is defined as the Cartesian product of a number of modes, some of which may be repeated. Thus, if the data consist of the measurements of certain characteristics of some entities, then the appropriate description is two-mode two-way data. If, however, the data are in the form of proximities between the entities, then they would be described as one-mode two-way data. The former is a more informative type of basic data set as it can be easily converted, if required, to the latter by suitable definition of a similarity or dissimilarity measure.

Consider such a two-mode two-way array, where $p$ attributes have been measured on each of $n$ entities, and where the aim is to partition these entities into $g$ groups, so that the entities within a group are, in some sense, homogeneous. If the partition were intended to correspond to $g$ externally existing groups, and there were data of known origin available from each of the groups for constructing estimates of the group densities, then a sample based allocation rule could be formed for assigning the $n$ entities to the $g$ possible groups. This discriminant analysis problem has been well studied, and the reader is directed to those books devoted to this topic, for example, Cacoullos (1973), Goldstein and Dillon (1978), Hand (1981), and Lachenbruch (1975), which have been supplemented recently by Choi (1986). There are also the relevant chapters in the rapidly growing list of text books on multivariate analysis, for example, Anderson (1984), Hawkins (1982), Kshirsagar (1972), Morrison (1976), and Seber (1984), to name a few. Another source of references is the pattern recognition literature. Devijver and Kittler (1982), Duda and Hart (1973), and Fukunaga (1972) are examples of texts on statistical pattern recognition. A single source for references in discriminant and cluster analyses and in pattern recognition is Volume 2 in the series Handbook of Statistics, edited by Krishnaiah and Kanal (1982).

In contrast to the discriminant analysis problem, in cluster analysis multivariate techniques are used to create groups amongst the entities, where there is no prior information regarding the underlying group structure, or at least where there are no available data from each of the groups if their existence is known. The need for cluster analysis has arisen in a natural way in many fields of study. For example, in taxonomy, clustering is used to identify subclasses of species. We shall be concentrating on another role of clustering where the partitioning of the data into relatively homogeneous groups is a data reduction exercise in its own right or part of an attempt to shed light on the phenomena of interest through a particular grouping.

In the last twenty years, the quantity of literature on cluster analysis has grown enormously, but unfortunately it has been mainly intradisciplinary. This lack of interdisciplinary communication has meant that large bodies of researchers appear to be unaware of one another (Anderberg, 1973). Various other attempts at classifying and reviewing cluster analysis methods have been made, including those by Cormack (1971), Clifford and Stephenson (1975), Everitt (1980), Gordon (1981), Hartigan (1975), Jardine and Sibson (1971), and Mezzich and Solomon (1980). To demonstrate the extent of the literature in this field, Seber (1984) reported that Sneath and Sokal (1973) contains approximately 1,600 references and Duran and Odell (1974) has 409 references. The various papers in Van Ryzin (1977) underscore the attention that cluster analysis has been receiving as an area of statistical methodological research.

Most clustering techniques are appropriate to data that are in the form of a two-mode two-way array ($p$ measurements on each of $n$ entities), or a one-mode two-way array (proximities measured between the $n$ entities), as described earlier. Also, they assume that the initial sample is unstructured, in the sense that there are no replications on any particular entity specifically identified as such, and that all the observation vectors on the entities are independent of one another. Within this framework available methods of seeking clusters can be categorized broadly as being hierarchical or nonhierarchical. The former class is one in which every cluster obtained at any stage is a merger or split of clusters at other stages. Thus it is possible to visualize not only the two extremes of clustering, that is, $n$ clusters with one entity per cluster (weak clustering) and a single cluster with all $n$ entities (strong clustering), but also a monotonically increasing strength of clustering as one goes from one level to another. A hierarchical strategy always optimizes a route between these two extremes (Williams, 1976). The route may be defined by progressive fusions, beginning with $n$ single entity groups and ending with a single group of $n$ entities (agglomerative hierarchy), or by progressive divisions, beginning with a single group and decomposing it into individual entities (divisive hierarchy).

In nonhierarchical procedures, new clusters are obtained by both lumping and splitting of old clusters, and although the two extremes of clustering are still the same, the intermediary stages of clustering do not have the natural monotone character of strength of clustering. Thus with a nonhierarchical strategy, it is the structure of the individual groups which is optimized, since these are made as homogeneous as possible (Williams, 1976). No route is defined between the groups and their constituent entities, so that the infrastructure of a group cannot be examined in this way. For those applications in which homogeneity of groups is of prime importance, the nonhierarchical strategies are very attractive. It will be seen in Section 1.12 on the work of Scott and Symons (1971) that many of the commonly used nonhierarchical clustering procedures correspond to applications of the likelihood approach for normal groups with various restrictions on the covariance matrices and with the indicator variables of group membership associated with the data treated as unknown parameters.

The mixture likelihood approach to clustering, which of course is nonhierarchical, has the potential to handle structured data because it is model based. The structure being referred to here is with respect to the collection and presentation of the data before a clustering technique is to be applied. It is not with reference to the underlying structure among the entities which the clustering technique is being used to identify. The structure of the data could be in the form of repeated observations on each entity by observing them in some experimental design, or it could be the representation of the information on the entities in a three-way array. Most clustering techniques assume the data are in the form of a one-mode or two-mode two-way array with no repeated observations as such. Hence the data have to be reduced to this form before a clustering technique can be applied. For example, consider a data set in the form of a three-mode three-way array, as representing the results of a large plant improvement program expressed as a genotype by attribute by environment matrix (Basford, 1982). This example, which is to be considered more fully in Chapter 7, is quite typical of experiments where various attributes are measured on each of a large number of genotypes grown in several environments. The aim of a cluster analysis on this set is to obtain a suitable grouping of the genotypes as a convenient labeling scheme, and to shed light on the underlying relationships between the genotypes. As stated above, most clustering techniques require the data in the form of a two-way array; here a genotype by attribute array, obtained by averaging over environments, or else a genotype by environment array for each attribute may be used. In the latter case a cluster analysis would have to be performed for each attribute of interest. Since, however, all the information collected is pertinent to the clustering of the genotypes, it would seem to be an advantage if a clustering technique were able to handle the entire three-way array in a single analysis. As demonstrated by Basford and McLachlan (1985c), the mixture likelihood approach can be applied to this set, directly incorporating the genotype by environment interaction, which is usually highly significant in large plant breeding trials.

## 1.4 MIXTURE LIKELIHOOD APPROACH TO CLUSTERING

In this section, we formulate likelihood estimation for finite mixture models, and then define the mixture likelihood approach to clustering, for data sets in the form of a two-mode two-way array. The extension to three-way data sets is given in Chapter 7. Multivariate observations on a set of $n$ entities forming a two-mode two-way array can be represented as $\mathbf{x}_1, \ldots, \mathbf{x}_n$, where each $\mathbf{x}_j$ is a vector of $p$-dimensions. Under the finite mixture models to be fitted here, each $\mathbf{x}_j$ can be viewed as arising from a superpopulation $G$ which is a mixture of a finite number, say $g$, of populations $G_1, \ldots, G_g$ in some proportions $\pi_1, \ldots, \pi_g$, respectively, where

$$\sum_{i=1}^{g} \pi_i = 1 \qquad \text{and} \qquad \pi_i \geq 0 \qquad (i = 1, \ldots, g).$$

The probability density function (p.d.f.) of an observation $\mathbf{x}$ in $G$ can therefore be represented in the finite mixture form,

$$f(\mathbf{x}; \phi) = \sum_{i=1}^{g} \pi_i f_i(\mathbf{x}; \theta), \tag{1.4.1}$$

where $f_i(\mathbf{x}; \theta)$ is the p.d.f. corresponding to $G_i$, and $\theta$ denotes the vector of all unknown parameters associated with the parametric forms adopted for these $g$ component densities. For example, in the particular case of multivariate normal component densities to be discussed in Section 2.1, $\theta$ consists of the elements of the mean vectors $\boldsymbol{\mu}_i$ and the distinct elements of the covariance matrices $\boldsymbol{\Sigma}_i$ for $i = 1, \ldots, g$. The vector

$$\phi = (\boldsymbol{\pi}', \theta')'$$

of all unknown parameters belongs to some parameter space $\boldsymbol{\Omega}$; the prime denotes vector transpose. Strictly speaking, since the elements of $\boldsymbol{\pi}$ sum to one, one of the mixing proportions $\pi_i$ in $\phi$ is redundant, but we shall not modify $\phi$ accordingly, at least explicitly. However, in our statements about the distribution of any estimator of $\phi$, it will be implicitly assumed that one of the $\pi_i$ has been deleted from $\phi$. Throughout this study we shall refer to the $G_i$ as either populations, components, or groups regardless of the underlying situation, although the use of the first term is really more appropriate to the case where there is *a priori* knowledge that the superpopulation is actually composed of $g$ distinct constituents, than to the usual case in cluster analysis where an artificial partition is to be imposed on the data.

In (1.4.1), the density functions are with respect to arbitrary measure on $\mathbf{R}^p$, so that $f_i(\mathbf{x};\theta)$ can be a mass function by the adoption of counting measure. In many applications, the component densities $f_i(\mathbf{x};\theta)$ are taken to belong to the same parametric family. We let $F(\mathbf{x};\phi)$ be the distribution function corresponding to the mixture density (1.4.1). Anderson (1979) called $F(\mathbf{x};\phi)$ a compound distribution, and reserved the use of the term mixture to distinguish between mixture and separate sampling schemes.

It is assumed for the present that $\mathbf{x}_1,\ldots,\mathbf{x}_n$ are the observed values of a random sample of size $n$ from $G$; that is, they are the realized values of $n$ independently and identically distributed (i.i.d.) random variables with common distribution function $F(\mathbf{x};\phi)$. This should be appropriate in many situations in practice. The assumption of independence is relaxed in Chapter 6, where the clustering of treatments from a randomized block design is considered. In this situation, each $\mathbf{x}_j$ can be viewed as recording the response of treatment $j$ in each of $p$ blocks, and so $\mathbf{x}_1,\ldots,\mathbf{x}_n$ are dependent if the block effects are taken to be random. Other situations not considered in this book, where the assumption of independence of the data may be inappropriate, occur in remote sensing. For instance, with the clustering of digitized images reported from the Landsat earth satellite for surface area elements, called pixels, the presence of a spatial correlation between neighboring pixels will cause the independence assumption to be violated; see, for example, Switzer (1983).

Our assumption about the present data set can be written as

$$\mathbf{x}_1,\ldots,\mathbf{x}_n \overset{iid}{\sim} F(\mathbf{x};\phi), \phi \in \boldsymbol{\Omega}. \tag{1.4.2}$$

Note that in (1.4.2) and the subsequent work random variables are not, for economy of notation, distinguished from their realizations by the use of the corresponding upper case letters, but in each instance the intended meaning is clear. The likelihood function for $\phi$ can be easily formed under (1.4.2), and an estimate of $\phi$ can be obtained as a solution of the likelihood equation,

$$\partial L(\phi)/\partial\phi = \mathbf{0},$$

where $L(\phi)$ denotes the log likelihood (the natural logarithm of the likelihood). We shall call our chosen solution $\hat{\phi}$ of the likelihood equation, the likelihood estimate of $\phi$. Unfortunately with mixture models, the likelihood equation usually has multiple roots. Their computation and the choice of root leading to $\hat{\phi}$ are to be discussed in the following sections of this chapter. But briefly, the aim of likelihood estimation is to determine $\hat{\phi}$ for each $n$ so that it defines a sequence of roots of the likelihood equation that is consistent and asymptotically efficient. Such a sequence is known to exist under suitable regularity conditions (Cramér, 1946). For estimation models in general, the likelihood usually has a global maximum in the interior of the parameter space. Then typically a sequence of roots of the likelihood equation with the desired asymptotic properties is provided by taking $\hat{\phi}$ for each $n$ to be the root which globally maximizes the likelihood; that is, $\hat{\phi}$ is the maximum likelihood estimate. However, for mixture models, examples can be easily found where the likelihood is unbounded, and so the maximum likelihood estimate does not exist. An example, first put forward by Kiefer and Wolfowitz (1956), is a mixture of two univariate normal densities with unknown different means and unknown different variances. In this example, the normal density can be replaced by any other density which is nonzero at the origin (Lehmann, 1983, page 442). It has been pointed out that if the inevitable discrete grouped nature of the data were taken into account, then these singularities would not occur; see, for example, Cox and Hinkley (1974, page 291).

Once $\hat{\phi}$ has been obtained, estimates of the posterior probabilities of population membership can be formed for each $\mathbf{x}_j$ (really the entity with observation $\mathbf{x}_j$) to give a probabilistic clustering. The posterior probability that $\mathbf{x}_j$ (again really the entity with observation $\mathbf{x}_j$, but we continue with this abuse of terminology for convenience) belongs to $G_i$ is given by

$$\begin{aligned}\tau_i(\mathbf{x}_j;\phi) &= \mathrm{pr}(j\text{th entity} \in G_i \mid \mathbf{x}_j;\phi) \\ &= \pi_i f_i(\mathbf{x}_j;\theta) \Big/ \sum_{t=1}^{g} \pi_t f_t(\mathbf{x}_j;\theta) \qquad (i=1,\ldots,g). \end{aligned} \tag{1.4.3}$$

A partitioning of $\mathbf{x}_1,\ldots,\mathbf{x}_n$ into $g$ nonoverlapping clusters can be effected by assigning each $\mathbf{x}_j$ to the population to which it has the highest estimated posterior probability of belonging; that is, to $G_t$ if

$$\tau_t(\mathbf{x}_j;\hat{\phi}) > \tau_i(\mathbf{x}_j;\hat{\phi}) \qquad (i=1,\ldots,g; \quad i \neq t). \tag{1.4.4}$$

For convenience, $\tau_i(\mathbf{x}_j;\hat{\phi})$ is denoted henceforth by $\hat{\tau}_{ij}$, and similarly, $\tau_i(\mathbf{x}_j;\phi)$ by $\tau_{ij}$. If $\phi$ were known, the allocation rule based on (1.4.4) would be the optimal or Bayes rule (Anderson, 1984, Chapter 6) which minimizes the overall error rate.

Before proceeding to discuss the computation of the likelihood estimate of $\phi$, the concept of identifiability is considered for finite mixture distributions.

## 1.5 IDENTIFIABILITY

In order to be able to estimate all the parameters in $\phi$ from $\mathbf{x}_1, \ldots, \mathbf{x}_n$, it is necessary that they should be identifiable. In general, the parametric family of p.d.f.'s $f(\mathbf{x}; \phi)$ is said to be identifiable if distinct values of $\phi$ determine distinct members of the family. This can be interpreted as follows in the case where $f(\mathbf{x}; \phi)$ defines a class of finite mixture densities according to (1.4.1). A class of finite mixtures is said to be identifiable for $\phi \in \mathbf{\Omega}$ if for any two members

$$f(\mathbf{x}; \phi) = \sum_{i=1}^{g} \pi_i f_i(\mathbf{x}; \theta),$$

and

$$f(\mathbf{x}; \phi^*) = \sum_{i=1}^{g^*} \pi_i^* f_i(\mathbf{x}; \theta^*),$$

then

$$f(\mathbf{x}; \phi) \equiv f(\mathbf{x}; \phi^*)$$

if and only if $g = g^*$ and we can permute the component labels so that

$$\pi_i = \pi_i^* \qquad \text{and} \qquad f_i(\mathbf{x}; \theta) \equiv f_i(\mathbf{x}; \theta^*) \qquad (i = 1, \ldots, g).$$

Here $\equiv$ implies equality of the densities for almost all $\mathbf{x}$ relative to the underlying measure on $\mathbf{R}^p$ appropriate for $f(\mathbf{x}; \phi)$. Titterington, Smith and Makov (1985, Section 3.1) have given a lucid account of the concept of identifiability for mixtures, including several examples and a survey of the literature on this topic.

It was tacitly assumed in the previous section that $\phi$ is identifiable. There is a difficulty with mixtures in that if $f_i(\mathbf{x}; \theta)$ and $f_j(\mathbf{x}; \theta)$ belong to the same parametric family, then $f(\mathbf{x}; \phi)$ will have the same value when the component labels $i$ and $j$ are interchanged in $\phi$. That is, although this class of mixtures may be identifiable, $\phi$ is not. Indeed, if all the $f_i(\mathbf{x}; \theta)$ belong to the same parametric family, then $f(\mathbf{x}; \phi)$ is invariant under the

$g!$ permutations of the component labels in $\phi$. In this case, if $\hat{\phi}$ corresponds to a local maximum, $L(\phi)$ will have at least $g!$ local maxima of the same value. This matter is to be discussed further in considering the consistency of the likelihood estimator of $\phi$ in Section 1.8.

However, this lack of identifiability of $\phi$ due to the interchanging of component labels is of no concern in practice, as it can be easily overcome by the imposition of an appropriate constraint on $\phi$. For example, the approach of Aitkin and Rubin (1985) in the case where all the components belong to the same parametric family, is to impose the restriction that

$$\pi_1 \geq \pi_2 \geq \cdots \geq \pi_g, \tag{1.5.1}$$

but to carry out likelihood estimation without this restriction. Finally, the component labels are permuted to achieve this restriction on the estimates of the mixing proportions. In the work to be presented here, we do not explicitly impose any restriction on the mixing proportions, but with any application of the mixture likelihood approach we report the result for only one of the possible arrangements of the elements of $\hat{\phi}$.

## 1.6 LIKELIHOOD ESTIMATION FOR MIXTURE MODELS VIA EM ALGORITHM

The computation of solutions of the likelihood equation under the finite mixture model (1.4.2) is described now for arbitrary component densities. This work is specialized to the important and commonly used model of normal component densities in Sections 2.1 and 2.2.

The likelihood equation for $\phi$, $\partial L(\phi)/\partial\phi = 0$, can be so manipulated that the likelihood estimate of $\phi$, $\hat{\phi}$, satisfies

$$\hat{\pi}_i = \sum_{j=1}^{n} \hat{\tau}_{ij}/n \qquad (i = 1, \ldots, g) \tag{1.6.1}$$

and

$$\sum_{i=1}^{g}\sum_{j=1}^{n} \hat{\tau}_{ij}\, \partial \log f_i(\mathbf{x}_j; \hat{\theta})/\partial\hat{\theta} = 0. \tag{1.6.2}$$

These manipulations were carried out by various researchers in the past in their efforts to solve the likelihood equation for mixture models with specific component densities; see, for example, Hasselblad (1966, 1969), Wolfe (1967, 1970), and Day (1969). They observed in their special cases that the equations (1.6.1) and (1.6.2) suggest an iterative computation of the solution which can be identified now with the direct application of the EM algorithm of Dempster, Laird and Rubin (1977); see Orchard and Woodbury (1972) for an example of earlier work on a general approach to likelihood estimation from incomplete data. On the use in particular instances of iterative schemes essentially equivalent to applications of the EM algorithm, Butler (1986) has pointed out examples in a mixture framework dating back to Newcomb (1886) and Jeffreys (1932), as described earlier in Section 1.2.

As one of the many examples in their general account on the use of the EM algorithm for the computation of likelihood estimates when the data can be viewed as incomplete, Dempster, Laird and Rubin (1977) considered the mixture problem. Corresponding to the formulation of finite mixture models in Section 1.4, we let the vector of indicator variables $\mathbf{z}_j = (z_{1j}, \ldots, z_{gj})'$ be defined by

$$z_{ij} = \begin{cases} 1, & \mathbf{x}_j \in G_i, \\ 0, & \mathbf{x}_j \notin G_i, \end{cases} \tag{1.6.3}$$

where $\mathbf{z}_1, \ldots, \mathbf{z}_n$ are independently and identically distributed according to a multinomial distribution consisting of one draw on $g$ categories with probabilities $\pi_1, \ldots, \pi_g$ respectively. We write

$$\mathbf{z}_1, \ldots, \mathbf{z}_n \overset{iid}{\sim} \text{Mult}_g(1, \pi). \tag{1.6.4}$$

In accordance with the mixture model (1.4.2), it is further assumed that $\mathbf{x}_1, \ldots, \mathbf{x}_n$ given $\mathbf{z}_1, \ldots, \mathbf{z}_n$ respectively are conditionally independent, and $\mathbf{x}_j$ given $\mathbf{z}_j$ has log density

$$\sum_{i=1}^{g} z_{ij} \log f_i(\mathbf{x}_j; \theta) \qquad (j = 1, \ldots, n).$$

Hence the log likelihood for the complete data,

$$\mathbf{X} = (\mathbf{x}_1', \ldots, \mathbf{x}_n')' \qquad \text{and} \qquad \mathbf{Z} = (\mathbf{z}_1', \ldots, \mathbf{z}_n')',$$

is given by

$$L_C(\phi) = \sum_{i=1}^{g} \sum_{j=1}^{n} z_{ij} \left\{ \log \pi_i + \log f_i(\mathbf{x}_j; \theta) \right\}. \tag{1.6.5}$$

The EM algorithm is applied to the mixture model (1.4.2) by treating $\mathbf{Z}$ as missing data. It is easy to program and it proceeds iteratively in two steps, E (for expectation) and M (for maximization). Using some initial value for $\phi$, say $\phi^{(0)}$, the E step requires the calculation of

$$Q(\phi, \phi^{(0)}) = E\{L_C(\phi) \mid \mathbf{X}; \phi^{(0)}\},$$

the expectation of the complete data log likelihood, $L_C(\phi)$, conditional on the observed data and the initial fit $\phi^{(0)}$ for $\phi$. It can be easily seen from (1.6.5), that this step is effected here simply by replacing each indicator variable $z_{ij}$ by its expectation conditional on $\mathbf{x}_j$, given by

$$E(z_{ij} \mid \mathbf{x}_j; \phi^{(0)}) = \tau_i(\mathbf{x}_j; \phi^{(0)}) \qquad (i = 1, \ldots, g). \tag{1.6.6}$$

That is, $z_{ij}$ is replaced by the initial estimate of the posterior probability that $\mathbf{x}_j$ belongs to $G_i$ $(i = 1, \ldots, g; j = 1, \ldots, n)$.

On the M step first time through, the intent is to choose the value of $\phi$, say $\phi^{(1)}$, that maximizes $Q(\phi, \phi^{(0)})$ which, from the E step, is equal here to $L_C(\phi)$ with each $z_{ij}$ replaced by $\tau_i(\mathbf{x}_j; \phi^{(0)})$. It leads therefore to solving equations (1.6.1) and (1.6.2) with $\hat{\tau}_{ij}$ replaced by $\tau_i(\mathbf{x}_j; \phi^{(0)})$. One nice feature of the EM algorithm is that the solution to the M step often exists in closed form, as is to be demonstrated for mixtures of normals in Section 2.1.

The E and M steps are alternated repeatedly, where in their subsequent executions, the initial fit $\phi^{(0)}$ is replaced by the current fit for $\phi$, say $\phi^{(k-1)}$ at the $k$th stage. Another nice feature of the EM algorithm is that the log likelihood for the incomplete data specification can never be decreased after an EM sequence. Hence the log likelihood under the mixture model satisfies

$$L(\phi^{(k+1)}) \geq L(\phi^{(k)}), \tag{1.6.7}$$

which implies that $L(\phi^{(k)})$ converges to some $L^*$ for a sequence bounded above. Dempster, Laird and Rubin (1977) showed that if the very weak condition that $Q(\phi, \psi)$ is continuous in both $\phi$ and $\psi$ holds, then $L^*$ will be a local maximum of $L(\phi)$, provided the sequence is not trapped at some saddle point. In which case, a small random perturbation of $\phi$ away from the saddle point will cause the EM algorithm to diverge from the saddle point. The condition that $Q(\phi, \psi)$ be continuous in both $\phi$ and $\psi$ should hold in most practical situations, and it has been shown that it is satisfied for component densities belonging to the important curved exponential family. The reader is referred to the detailed account of the convergence properties of the EM algorithm in a general setting given by Wu (1983) who addressed, in particular, the problem that the convergence of $L(\phi^{(k)})$ to $L^*$ does not automatically imply the convergence of $\phi^{(k)}$ to a point $\phi^*$. On this same matter, Boyles (1983) presented an example of a generalized EM sequence which converges to the circle of unit radius and not to a single point.

Unfortunately, convergence with the EM algorithm is generally quite slow and it does not yield directly the observed information matrix which may be needed for the estimation of the standard errors of the elements of $\hat{\phi}$. However, Louis (1982) has devised a procedure for extracting the observed information matrix when using the EM algorithm, as well as developing a method for speeding up its convergence. The extraction of the observed information matrix is to be described in Section 1.9.

Concerning other possible alternatives to the EM algorithm, the Newton-Raphson method may not increase the likelihood after each iteration, although when it does converge the convergence can be quite rapid, in particular, when the log likelihood is well approximated by a quadratic function. As a Newton iterative scheme requires the inversion of the matrix of second derivatives of the log likelihood, it automatically provides an estimate of the covariance matrix of $\hat{\phi}$, although this inversion may not be computationally convenient at each step if the dimension of $\phi$ is high. For a mixture of two univariate normal densities, Everitt (1984a) has compared the EM algorithm with five other methods for finding the likelihood estimates. Titterington (1984a) has considered recursive estimation for a mixture of two univariate normal densities as part of a wider study of recursive estimation for incomplete data and its link with the EM algorithm. More recently, Hathaway (1986a) showed that the EM algorithm for mixture problems can be interpreted as a method of coordinate descent on a particular objective function.

The Appendix contains the FORTRAN listing of computer programs for likelihood estimation via the EM algorithm for mixtures of normal distributions. Agha and Ibrahim (1984) have provided an ALGOL 60 listing of this algorithm for likelihood estimation for univariate mixtures of binomial, Poisson, exponential and normal distributions. Also, as noted by Aitkin (1980a), the iterative computations required for the EM algorithm in the fitting of mixture models are particularly simple in GLIM.

## 1.7 STARTING VALUES FOR EM ALGORITHM

As explained in the previous section, the EM algorithm is started from some initial value of $\phi$, $\phi^{(0)}$. We can either give a value for $\phi^{(0)}$, or initially partition the data into the specified number of groups $g$ according, perhaps, to some ad hoc criterion, and then take $\phi^{(0)}$ to be the estimate of $\phi$ based on this partition as if it represented a true grouping of the data. That is, we execute the M step of the EM algorithm with the initial estimate of the posterior probability, $\tau_i(\mathbf{x}_j; \phi^{(0)})$, set equal to 1 or 0 according as to whether $\mathbf{x}_j$ lies in $G_i$ or not under this partition. For example, one ad hoc way of initially partitioning the data in the case of, say $g = 2$ normal populations with the same covariance matrices, would be to plot the data for selections of two of the $p$ variables, and then draw a line which divides the bivariate data into two groups which have a scatter that appears normal; see, for example, O'Neill (1978) and Ganesalingam and McLachlan (1979b). On the selection of a reduced set of variables from the full observation vector, note that Chang (1983) has shown that the practice of applying principal components to reduce the dimension of the data before clustering is not justified in general. By considering a mixture of two normal distributions with a common covariance matrix, it was shown that the components with the large eigenvalues do not necessarily provide more information (distance) between the groups.

As commented in the last section, convergence with the EM algorithm is very slow, and the situation will be exacerbated by a poor choice of $\phi^{(0)}$. Indeed, in some cases where the likelihood is unbounded on the edge of the parameter space, the sequence of estimates $\{\phi^{(k)}\}$ generated by the EM algorithm may diverge if $\phi^{(0)}$ is chosen too close to the boundary. This matter is discussed further in Section 2.1. Another problem with mixture models is that the likelihood equation will usually have multiple roots, and so the EM algorithm should be applied from a wide choice of starting values in any search for all local maxima.

It will be illustrated in Section 3.2 that a key factor in obtaining a satisfactory clustering of the data at hand is the accuracy of the estimate of $\boldsymbol{\pi}$, the vector of mixing proportions. For univariate mixtures Fowlkes (1979) suggested the determination of the point of inflexion in quantile-quantile ($Q$-$Q$) plots to estimate first the mixing proportions of the underlying populations. The remaining parameters can then be estimated from the sample partitioned into groups in accordance with the estimates of the mixing proportions. After the completion of a search for the various local maxima, there is the problem of which root of the likelihood equation to use. If there is available some consistent estimator of $\phi$, then the estimator defined by the root of the likelihood equation closest to it is also consistent, and hence efficient (Lehmann, 1983, page 421), providing regularity conditions hold for the existence of a sequence of consistent and asymptotically efficient solutions of the likelihood equation. Where computationally feasible, the method of moments as discussed in Section 1.2 provides a way of constructing a $\sqrt{n}$-consistent estimator of $\phi$. Quandt and Ramsey (1978) proposed a method for obtaining a $\sqrt{n}$-consistent estimator of $\phi$, in terms of minimizing a distance function based on the moment generating function, but there can be numerical complexities; see the discussion following that paper. Minimum distance estimators of this type are to be discussed further in Section 4.6. The usefulness of a $\sqrt{n}$-consistent estimator of $\phi$, say $\tilde{\phi}$, is that under the same regularity conditions referred to above, a consistent and asymptotically efficient estimator of $\phi$ is obtained by a one-step approximation to the solution of the likelihood equation according to a Newton iterative scheme started from $\tilde{\phi}$. This approach does not require the determination of the closest root of the likelihood equation to $\tilde{\phi}$, but its implementation does require the inversion of a symmetric matrix whose dimension is equal to the total of unknown parameters. Hence it may not be as convenient computationally as the EM algorithm applied from $\tilde{\phi}$.

In the absence of the observed value of any known consistent estimator of $\phi$, an obvious choice for the root of the likelihood equation is the one corresponding to the largest of the local maxima (assuming all have been located), although it does not necessarily follow that the consequent estimator is consistent and asymptotically efficient (Lehmann, 1980, page 234). However, at least for mixtures of univariate normal distributions, the estimator of $\phi$ corresponding to the largest of the local maxima is consistent and efficient. The properties of this estimator for mixtures of multivariate normal distributions are to be examined in Section 2.1.

## 1.8 PROPERTIES OF LIKELIHOOD ESTIMATORS FOR MIXTURE MODELS

If the maximum likelihood estimator of $\phi$ exists for a mixture distribution with compact parameter space, then the conditions of Wald (1949) ensure that it is strongly consistent; see Redner (1981). This is assuming that $\phi$ is identifiable. It has been seen in Section 1.5 that, although a class of mixture densities may be identifiable, $\phi$ may not. Let $\phi_0$ denote the true value of $\phi$ and let $\Omega_0$ denote the subset of the parameter space for which

$$f(\mathbf{x};\phi) = f(\mathbf{x};\phi_0)$$

for almost all $\mathbf{x}$ in $\mathbf{R}^p$. This set will contain more than the single point $\phi_0$ in the case of component densities belonging to the same parametric family. For then, as discussed in Section 1.5, a permutation of the component labels in $\phi$ does not alter the value of $f(\mathbf{x};\phi)$. This particular identifiability problem can be avoided by the imposition of a constraint on $\phi$ like (1.5.1).

Redner's (1981) work, however, was specifically aimed at families of distributions which are not identifiable. His results imply for mixture families with compact parameter space that, under the conditions of Wald (1949) except for the identifiability of $\phi$, $L(\phi)$ is almost surely maximized in a neighborhood of $\Omega_0$. More precisely, Redner (1981) referred to it as convergence of the maximum likelihood estimator in the topology of the quotient space obtained by collapsing $\Omega_0$ into a single point; see Ghosh and Sen (1985), Li and Sedransk (1986), and Redner and Walker (1984) for further discussion. In the sequel it is implicitly assumed that $\Omega_0$ is a singleton set; that is, $\phi$ is identifiable or, in the terminology of Ghosh and Sen (1985), the class of mixture densities is strongly identifiable.

As noted earlier, in some instances the log likelihood $L(\phi)$ under the mixture model (1.4.2) is unbounded, and so the maximum likelihood estimator of $\phi$ does not exist. However, as emphasized recently by Lehmann (1980, 1983), the essential aim of likelihood estimation is the determination of a sequence of roots of the likelihood equation that is consistent and asymptotically efficient. Under suitable regularity conditions, it is well known (Cramér, 1946) that there will exist such a sequence of roots. With probability tending to one, these roots correspond to local maxima in the interior of the parameter space. As stressed by Lehmann (1980, 1983), these roots do not necessarily correspond to global maxima. But usually this sequence of roots with the desirable asymptotic properties will correspond to global maxima, or if the latter do not exist, to the largest of the local maxima located in the interior of the parameter space. Whether this is the case for mixtures of normal distributions will be discussed in Section 2.1, where available results for this specific parametric family are reviewed.

Concerning likelihood estimation for mixtures in general, Redner and Walker (1984) and Peters and Walker (1978) have given, for the class of identifiable mixtures, the regularity conditions which must be satisfied in order for there to exist a sequence of roots of the likelihood equation $\partial L(\phi)/\partial\phi = \mathbf{0}$ with the properties of consistency, efficiency and asymptotic normality. The form of the conditions, which are essentially multivariate generalizations of Cramér's (1946) results for the corresponding properties in a general context, suggest that they should hold for many parametric families.

## 1.9 INFORMATION MATRIX FOR MIXTURE MODELS

We now consider the amount of information in the sample $\mathbf{x}_1, \ldots, \mathbf{x}_n$ under the mixture model (1.4.2). Let $\mathbf{I}(\phi)$ be the matrix defined by

$$\mathbf{I}(\phi) = ((-\partial^2 L(\phi)/\partial\phi_i \partial\phi_j)). \tag{1.9.1}$$

In a general context, Efron and Hinkley (1978) have termed $\mathbf{I}(\hat{\phi})$ as the observed Fisher information matrix as distinguished from the expected Fisher information matrix, $\mathcal{I}(\phi)$, given by the expectation of (1.9.1). For one-parameter families they presented a frequentist justification for preferring $1/I(\hat{\phi})$ to $1/\mathcal{I}(\hat{\phi})$ as an estimate of the variance of the likelihood estimator $\hat{\phi}$.

For the mixture model the observed information matrix $\mathbf{I}(\hat{\phi})$ is easier to compute than $\mathcal{I}(\hat{\phi})$, as Louis (1982) showed that $\mathbf{I}(\hat{\phi})$ can be expressed in terms of the gradient and curvature of the log likelihood function $L_C(\phi)$ for the complete data $\mathbf{X}$ and $\mathbf{Z}$. This applies to any use of the EM algorithm and not just to mixture models. It was shown that $\mathbf{I}(\hat{\phi})$ is equal to the value at $\phi = \hat{\phi}$ of

$$E[\{\mathbf{I}_1(\phi) - \mathbf{I}_2(\phi)\} \mid \mathbf{X}; \phi], \tag{1.9.2}$$

where

$$\mathbf{I}_1(\phi) = ((-\partial^2 L_C(\phi)/\partial\phi_i \partial\phi_j))$$

and

$$\mathbf{I}_2(\phi) = ((\{\partial L_C(\phi)/\partial\phi_i\}\{\partial L_C(\phi)/\partial\phi_j\})).$$

The log likelihood for the complete data, $L_C(\phi)$ can be written as

$$L_C(\phi) = \sum_{j=1}^{n} L_C(\phi; \mathbf{x}_j, \mathbf{z}_j),$$

where

$$L_C(\phi; \mathbf{x}_j, \mathbf{z}_j) = \sum_{i=1}^{g} z_{ij}\{\log \pi_i + \log f_i(\mathbf{x}_j; \theta)\}$$

is the log likelihood for the single complete data point $(\mathbf{x}_j', \mathbf{z}_j')'$. For independent data as assumed here, it follows that (1.9.2) evaluated at $\phi = \hat{\phi}$ can be reduced to give

$$\mathbf{I}(\hat{\phi}) \approx \sum_{j=1}^{n} \hat{\mathbf{h}}_j \hat{\mathbf{h}}_j', \tag{1.9.3}$$

where

$$\mathbf{h}(\phi; \mathbf{x}_j, \mathbf{z}_j) = \partial L_C(\phi; \mathbf{x}_j, \mathbf{z}_j)/\partial\phi$$

and where

$$\hat{\mathbf{h}}_j = \mathbf{h}(\hat{\phi}; \mathbf{x}_j, \hat{\boldsymbol{\tau}}_j)$$

and

$$\hat{\boldsymbol{\tau}}_j = (\hat{\tau}_{1j}, \ldots, \hat{\tau}_{gj})'$$

for $j = 1, \ldots, n$. Hence $\mathbf{I}(\hat{\phi})$ is approximated solely in terms of the gradient vector of $L_C(\phi; \mathbf{x}_j, \mathbf{z}_j)$ at $\phi = \hat{\phi}$, where the unknown vector $\mathbf{z}_j$ of indicator variables is replaced by the vector $\hat{\boldsymbol{\tau}}_j$ of the estimated posterior probabilities of group membership for $\mathbf{x}_j$ $(j = 1, \ldots, n)$. If only the mixing proportions are unknown, (1.9.3) holds exactly. The computation of $\mathbf{I}(\hat{\phi})$ from (1.9.3) for normal component distributions will be discussed in Section 2.4.

## 1.10 TESTS FOR THE NUMBER OF COMPONENTS IN A MIXTURE

Testing for the number of components $g$ in a mixture is an important but very difficult problem which has not been completely resolved. For example, with applications of mixture models in cluster or latent class analyses, the problem arises with the question of how many clusters or latent classes there are. An obvious way of approaching the problem is to use the likelihood ratio test statistic $\lambda$ to test for the smallest value of $g$ compatible with the data.

Unfortunately with mixture models, regularity conditions do not hold for $-2\log\lambda$ to have its usual asymptotic null distribution of chi-squared with degrees of freedom equal to the difference in the number of parameters in the two hypotheses. To briefly illustrate why this is so, consider a mixture of two univariate normal densities; that is, $p = 1$ in the model (1.4.1). The null hypothesis that there is one underlying population,

$$H_0 : g = 1, \tag{1.10.1}$$

can be approached by testing whether $\pi_1 = 1$, which is on the boundary of the parameter space with a consequent breakdown in the standard regularity conditions. Alternatively, we can view $H_0$ as testing for whether $\mu_1 = \mu_2$ and $\sigma_1^2 = \sigma_2^2$, where now the value of $\pi_1$ is irrelevant. If for a specified value of $\pi_1$ regularity conditions held, so that $-2\log\lambda$ under $H_0$ were distributed asymptotically as chi-squared, then the null distribution of $-2\log\lambda$ where $\pi_1$ is unspecified, would correspond to the maximum of a set of dependent chi-squared variables. However, even for specified mixing proportions the usual regularity conditions do not hold; see Quinn and McLachlan (1986).

The recent article by Ghosh and Sen (1985) provides a comprehensive account of the breakdown in regularity conditions for the classical asymptotic theory to hold for the likelihood ratio test. Other relevant references in addition to those to be discussed below include Bock (1985), Hartigan (1977), and Moran (1973).

For a mixture of two known but general univariate densities in unknown proportions, Titterington (1981) and Titterington, Smith and Makov (1985) considered the likelihood ratio test of $H_0 : g = 1$ $(\pi_1 = 1)$ versus $H_1 : g = 2$ $(\pi_1 < 1)$. They showed asymptotically under $H_0$ that $-2\log\lambda$ is zero with probability 0.5 and, with the same probability, is distributed as chi-squared with one degree of freedom. Another way of expressing this is that the asymptotic null distribution of $-2\log\lambda$ is the same as the distribution of

$$\{\max(0, Y)\}^2,$$

where $Y$ is a standard normal random variable.

Hartigan (1985a,b) obtained the same result for the asymptotic null distribution of $-2\log\lambda$ in the case of a mixture in unknown proportions of two univariate normal densities with known common variance and known means $\mu_1$ and $\mu_2$ where, as in the previous example, the null hypothesis $H_0 : g = 1$ was specified by $\pi_1 = 1$. This example was considered also by Ghosh and Sen (1985) in the course of their development of asymptotic theory for the distribution of the likelihood ratio test statistic for mixture models. They were able to derive the limiting null distribution of $-2\log\lambda$ for unknown but identifiable $\mu_1$ and $\mu_2$, where $\mu_2$ lies in a compact set. They showed in the limit, that $-2\log\lambda$ is distributed as a certain functional,

$$\left[\max\left\{0, \sup_{\mu_2} Y(\mu_2)\right\}\right]^2,$$

where $Y(\cdot)$ is a Gaussian process with zero mean and covariance kernel depending on the true value of $\mu_1$ under $H_0$, and the variance of $Y(\mu_2)$ is unity for all $\mu_2$. Ghosh and Sen (1985) established a similar result for component densities from a general parametric family under certain conditions. For the case where the vector of parameters $\phi$ was not assumed to be identifiable, they imposed a separation condition on the values of $\phi$ under $H_0$ and $H_1$.

The above result of Hartigan (1985a,b) was derived as a preliminary step in his argument that if $\mu_2$ is unknown with no restrictions on it, then $-2 \log \lambda$ is asymptotically infinite. He conjectured that if $\mu_2$ were given a prior distribution, then the resulting test may have better asymptotic behavior. This approach was advocated too by Ghosh and Sen (1985), and they suggested that the prior distribution should probably be chosen so that the associated test is asymptotically locally minimax. As a first step in this direction, they noted for their normal mixture example that the likelihood ratio test is not asymptotically locally minimax, but it is if $\mu_2$ were known; that is, if a degenerate prior were placed on $\mu_2$. A similar result was proved for general exponential component densities. Li and Sedransk (1986) have considered recently a Bayesian approach to testing for the number of components in a mixture.

One of the initial attempts on this problem was by Wolfe (1971). He considered the likelihood ratio test for assessing the null hypothesis that the data arise from a mixture of $g_1$ populations versus the alternative that they arise from $g_2$ ($g_1 < g_2$). He put forward a recommendation on the basis of a small scale simulation study performed for $g_1 = 1$ and $g_2 = 2$. It follows from his proposal that the null distribution of $2 \log \lambda$ would be approximated as

$$-2c \log \lambda \sim \chi^2_d, \tag{1.10.2}$$

where the degrees of freedom, $d$, is taken to be twice the difference in the number of parameters in the two hypotheses, not including the mixing proportions. His suggested value of $c$ is

$$\left(n - 1 - p - \tfrac{1}{2} g_2\right) / n,$$

which is similar to the correction factor

$$\left\{n - 1 - \tfrac{1}{2}(p + g_2)\right\} / n \tag{1.10.3}$$

derived by Bartlett (1938), for improving the asymptotic chi-squared distribution of $-2 \log \lambda$ for the problem of testing the equality of $g_2$ means in a multivariate analysis of variance. It applies to the case of only one observation on each entity to be clustered; that is, $r = 1$. When there are $r$ replications per entity, as in the clustering of treatments in a randomized complete block design (RCBD), the corresponding value of $c$ would be

$$\left\{r(n-1) - \tfrac{1}{2}(p+g_2)\right\} / nr.$$

It is analogous to the value of the correction factor obtained by Kendall (1965, page 134) by first eliminating the block differences from the variation before using Bartlett's approximation for testing between $g_2$ means in a multivariate analysis of variance.

On the basis of some simulations for normal populations, Hernandez-Avila (1979) concluded it was reasonable to work with the suggestion (1.10.2) of Wolfe (1971). Also, some recent simulations performed by Everitt (1981b) for testing $g = 1$ versus $g = 2$ normal populations with equal covariance matrices suggest that $-2c \log \lambda$ with $c$ defined by (1.10.3) is very well approximated by a chi-squared distribution under (1.10.1). However, in these simulations the sample size $n$ was equal to 50 and the ratio of $n$ to the number of dimensions $p$ was not less than five. The more extensive simulations of Everitt (1981a) suggest that this ratio $n/p$ needs to be at least five if Wolfe's approximation is to be applicable for the determination of $P$-values. Even then the simulated power of the test was low in cases where the Mahalanobis distance $\Delta$ was not greater than two.

Hence it is recommended in general that the outcome of Wolfe's modified likelihood ratio test should not be rigidly interpreted, but rather used as a guide to the possible number of underlying groups. In the numerical examples to be presented later our use of (1.10.2) will be in this spirit. The constant $c$ will be taken to be unity since we are only after a rough guide. Also, there is no theoretical justification for taking the null distribution of $-2 \log \lambda$ to be chi-squared in the first place, let alone for its subsequent refinement through the use of the multiplicative constant $c$ by analogy to results in the regular case. It is suggested that use be made also of the estimates of posterior probabilities of group membership in the choice of $g$. They can be examined for values of $g$ near to the value accepted according to the likelihood ratio test, and may therefore be of assistance in leading to a final decision as to the number of underlying groups. On another cautionary note on the use of the likelihood ratio for testing for the number of groups, it is sensitive to the normality assumption concerning the group densities. As reiterated recently by Hartigan and Hartigan (1985), this test may well reject the null hypothesis of $g = 1$ with a high probability in the case of a long tailed unimodal distribution, since the latter may resemble more closely a mixture of normal densities than a single normal. For example, the lognormal density can be well approximated by a mixture of two univariate normal densities with a common variance; see Titterington, Smith and Makov (1985, page 30).

Aitkin, Anderson and Hinde (1981) have reservations about the adequacy of chi-squared approximations like (1.10.2) for the null distribution of $-2\log\lambda$, and in the context of latent class models outlined a solution to the problem using essentially a bootstrap approach. The "bootstrap" was introduced by Efron (1979), who has investigated it further in a series of articles; see Efron (1982, 1983, 1984) and the references therein. It is a powerful technique which permits the variability in a random quantity to be assessed using just the data at hand. An estimate $\hat{F}$ of the underlying distribution is formed from the observed sample. Conditional on the latter, the sampling distribution of the random quantity of interest with $F$ replaced by $\hat{F}$, defines its so-called bootstrap distribution, which provides an approximation to its true distribution. It is assumed that $\hat{F}$ has been so formed that the stochastic structure of the model has been preserved. Usually, it is impossible to express the bootstrap distribution in simple form, and it must be approximated by Monte Carlo methods whereby pseudo-random samples (bootstrap samples) are drawn from $\hat{F}$. The bootstrap method can be implemented nonparametrically by using the empirical distribution function constructed from the original data. In the following application the bootstrap is applied in a parametric framework in which the bootstrap samples are drawn from the parametric likelihood estimate of the underlying distribution function.

The log likelihood ratio statistic for the test of the null hypothesis $H_0 : g = g_1$ groups versus the alternative $H_1 : g = g_2$ can be bootstrapped as follows. Proceeding under $H_0$, a bootstrap sample is generated from a mixture of $g_1$ groups where, in the specified form of their densities, unknown parameters are replaced by their likelihood estimates formed under $H_0$ from the original sample. The value of $-2\log\lambda$ is computed for the bootstrap sample after fitting mixture models for $g = g_1$ and $g_2$ in turn to it. This process is repeated independently a number of times $K$, and the replicated values of $-2\log\lambda$ formed from the successive bootstrap samples provide an assessment of the bootstrap, and hence of the true, null distribution of $-2\log\lambda$. It enables an approximation to be made to the achieved level of significance $P$ corresponding to the value of $-2\log\lambda$ evaluated from the original sample.

If a very accurate estimate of the $P$-value were required, then $K$ may have to be very large. Indeed for less complicated models than mixtures, Efron (1984), Efron and Tibshirani (1986), and Tibshirani (1985) have shown that whereas 50 to 100 bootstrap replications may be sufficient for standard error and bias estimation, a larger number, say 350, are needed to give a useful estimate of a percentile or $P$-value, and many more for a highly accurate assessment. Usually, however, there is no interest in estimating a $P$-value with high precision. Even with a limited replication number $K$, the amount of computation involved is still considerable, in particular for values of $g_1$ and $g_2$ not close to one.

In the narrower sense where the decision to be made concerns solely the rejection or retention of the null hypothesis at a specified significance level $\alpha$, Aitkin, Anderson and Hinde (1981) noted how analogous to the Monte Carlo test procedure of Hope (1968), the bootstrap replications can be used to provide a test of approximate size $\alpha$. The test which rejects $H_0$ if $-2\log\lambda$ for the original data is greater than the $j$th smallest of its $K$ bootstrap replications, has size

$$\alpha = 1 - j/(K+1) \tag{1.10.4}$$

approximately. For if any difference between the bootstrap and true null distributions of $-2\log\lambda$ is ignored, then the original and subsequent bootstrap values of $-2\log\lambda$ can be treated as the realizations of a random sample of size $K+1$, and the probability that a specified member is greater than $j$ of the others is $1 - j/(K+1)$. For some hypotheses the null distribution of $\lambda$ will not depend on any unknown parameters, and so then there will be no difference between the bootstrap and true null distributions of $-2\log\lambda$. An example is the case of normal populations with all parameters unknown where $g_1 = 1$ under $H_0$. The normality assumption is not crucial in this example.

Note that the result (1.10.4) applies to the unconditional size of the test and not to its size conditional on the $K$ bootstrap values of $-2\log\lambda$. For a specified significance level $\alpha$, the values of $j$ and $K$ can be appropriately chosen according to (1.10.4). For example, for $\alpha = 0.05$, the smallest value of $K$ needed is 19 with $j = 19$. As cautioned above on the estimation of the $P$-value for the likelihood ratio test, $K$ needs to be very large to ensure an accurate assessment. In the present context the size of $K$ manifests itself in the power of the test; see Hope (1968). Although the test may have essentially the prescribed size for small $K$, its power may be well below its limiting value as $K \to \infty$. For the 0.05 level test of a single normal population versus a mixture of two normal homoscedastic populations, McLachlan (1986b) performed some simulations to demonstrate the improvement in the power as $K$ is increased from 19 through 39 to 99.

In an attempt to overcome some of the problems with the use of the likelihood ratio statistic in testing for the number of groups completely within a frequentist framework, Aitkin and Rubin (1985) adopted an approach which places a prior distribution on the vector of mixing proportions $\boldsymbol{\pi}$. Likelihood estimation is undertaken for the remaining parameters in $\boldsymbol{\phi}$, $\boldsymbol{\theta}$, after first integrating the likelihood of $\boldsymbol{\phi}$ over the prior distribution of $\boldsymbol{\pi}$, $f_{\pi}(\boldsymbol{\pi})$. In this account of their proposal we shall write $L(\boldsymbol{\phi})$ as $L(\boldsymbol{\pi}, \boldsymbol{\theta})$, as we wish to let $L(\theta)$ denote the log likelihood given by

$$L(\theta) = \log\left[\int \exp\{L(\pi, \theta)\} f_\pi(\pi)\, d\pi\right].$$

The likelihood estimate of $\theta$ using $L(\theta)$ is denoted by $\tilde{\theta}$ and, as before, $\hat{\phi} = (\hat{\pi}', \hat{\theta}')'$ denotes the likelihood estimate of $\phi$ based on $L(\pi, \theta)$.

Aitkin and Rubin (1985) noted that $\tilde{\theta}$ can be obtained using the EM algorithm with the vector of indicator variables, $\mathbf{Z} = (\mathbf{z}_1', \ldots, \mathbf{z}_n')'$, and $\pi$ treated as unknown. It follows that $\tilde{\theta}$ can be computed iteratively in the same manner as $\hat{\theta}$ from (1.6.2). Whereas at the $(k+1)$th stage of the $E$ step in the computation of $\hat{\theta}$, $z_{ij}$ is replaced by the posterior probability of $\mathbf{x}_j$ using the current estimate $\phi^{(k)}$ for $\phi$, $\tau_i(\mathbf{x}_j; \pi^{(k)}, \theta^{(k)})$, in the computation of $\tilde{\theta}$, $z_{ij}$ is replaced by

$$\int \tau_i(\mathbf{x}_j; \pi, \theta^{(k)}) f_\pi(\pi \mid \mathbf{X}, \theta^{(k)})\, d\pi, \tag{1.10.5}$$

where the posterior density of $\pi$ given $\mathbf{X}$ and $\theta = \theta^{(k)}$ is proportional to

$$f_\pi(\pi \mid \mathbf{X}, \theta^{(k)}) \propto f_\pi(\pi) \exp\{L(\theta^{(k)})\}.$$

It can be seen with this proposal that there is the need to perform a numerical integration (1.10.5). However, according to Aitkin and Rubin (1985), $\tilde{\theta}$ is often quite close to $\hat{\theta}$, which can be used for $\theta$ in initially forming the posterior probabilities from (1.10.5). Likelihood ratio tests for intermediate numbers of groups, for example, $H_0 : g = 2$ versus $H_1 : g = 3$, are not recommended since in general the $g = 2$ model is not nested in the $g = 3$ model.

An advantage of this proposal is that any null hypothesis about the number of groups is specified in the interior of the parameter space of $\theta$, and so there are not some of the problems as in the formulation using $\phi$. However, it follows from the recent results of Quinn, McLachlan, and Hjort (1987) that the asymptotic null distribution of $-2\log\lambda$ will not necessarily be chi-squared, as regularity conditions still do not hold. In particular, for the test of $H_0 : g = 1$ versus $H_1 : g = g_2$ ($> 1$) for component densities belonging to the same parametric family, they showed that $1/n$ times the observed information matrix has negative eigenvalues with nonzero probability under $H_0$, as $n \to \infty$.

Sclove (1983) and Bozdogan and Sclove (1984) have proposed using Akaike's information criterion (AIC) for the choice of the number of groups

$g$ in clustering. Their proposal in the context of the mixture model (1.4.2) is to choose the $g$ which minimizes

$$\mathrm{AIC}(g) = -2L(\hat{\phi}) + 2N(g),$$

where $N(g)$ is the number of free parameters within the model. However, as explained by Titterington, Smith and Makov (1985), this criterion relies essentially on the same regularity conditions needed for $-2\log\lambda$ to have its usual asymptotic distribution under the null hypothesis.

Another way of approaching the problem of the number of components in a mixture is through an assessment of the number of modes. Care, however, must be exercised as bimodality can occur in a sample from a single normal population. On the other hand, a mixture distribution can be unimodal; for example, a mixture in equal proportions of univariate normal densities with means $\mu_1$ and $\mu_2$ and variance $\sigma^2$ is bimodal if and only if $|\mu_1-\mu_2|/\sigma > 2$. Conditions for a mixture of two univariate normal densities to be bimodal are discussed by Everitt and Hand (1981, Section 2.2) and also by Titterington, Smith and Makov (1985, Section 5.5) who included the multivariate case. Inferential procedures for assessing the number of modes of a distribution are described by Titterington, Smith and Makov (1985, Section 5.6), including the univariate technique of Silverman (1981, 1983) which uses the kernel method to estimate the density function nonparametrically and which permits the assessment of modality under certain circumstances. More recently, Wong (1985) showed in the univariate case that a bootstrap procedure based on the $k$th nearest neighbor clustering method of Wong and Lane (1983), is an effective tool for determining the number of populations under the assumption that a population corresponds to a mode in the mixture density. His statistic is similar to that of Silverman (1981) for kernel density estimates.

On this problem of assessing the number of components in a mixture, there have also been various attempts made in special cases. For instance, Anderson (1985) considered this problem in association with the mixture likelihood approach for component densities which were univariate normal with a known common variance. Using the score statistic $\partial L(\boldsymbol{\pi})/\partial\pi_1$ for a mixture of two known univariate distributions, Durairajan and Kale (1979) developed the locally most powerful test of the null hypothesis $\pi_1 = 1$ against the alternative $\pi_1 < 1$. Durairajan and Kale (1982) subsequently gave the locally most powerful similar test in the case where the two component densities were known up to a single nuisance parameter common to both. For a mixture of known univariate distributions, Henna (1985) proposed that the number of components be estimated by first choosing $\boldsymbol{\pi}$ to minimize the average squared distance between the mixture distribution function $F(x; \boldsymbol{\pi})$ and its empirical version,

$$\delta(\boldsymbol{\pi}) = \sum_{j=1}^{n} \{F(x_{(j)}; \boldsymbol{\pi}) - j/n\}^2/n. \qquad (1.10.6)$$

where $x_{(j)}$ denotes the $j$th order statistic ($j = 1, \ldots, n$). If $\hat{\boldsymbol{\pi}}_{\mathrm{MD}}$ denotes the estimate of $\boldsymbol{\pi}$ so obtained, then the number of components is estimated by $\hat{g}$, the minimal integer $g$ such that

$$\delta(\hat{\boldsymbol{\pi}}_{\mathrm{MD}}) < \Psi^2(n)/n,$$

where $\Psi(n) \uparrow \infty$ and $\Psi^2(n)/n \to 0$, as $n \to \infty$, and

$$\sum_{n=1}^{\infty} \{\Psi^2(n)/n\} \exp\{-2\Psi^2(n)\} < \infty.$$

For an identifiable class of finite mixtures, Henna (1985) established that $\hat{g}$ converges to the true value of $g$, $g_0$, with probability one as $n \to \infty$, providing $0 < \pi_i < 1$ ($i = 1, \ldots, g_0$). An account of the estimation of mixing proportions through the use of (1.10.6) and other distance measures is to be given in Section 4.6.

## 1.11 PARTIAL CLASSIFICATION OF THE DATA

We now consider the situation where the mixture model is not just an artificial device by which to effect a clustering of the unclassified data $\mathbf{x}_1, \ldots, \mathbf{x}_n$, but where the superpopulation $G$ is a genuine mixture of $g$ populations $G_1, \ldots, G_g$ in proportions $\pi_1, \ldots, \pi_g$, respectively. It is assumed that in addition to the unclassified data $\mathbf{x}_1, \ldots, \mathbf{x}_n$ taken from $G$, there are also available observations $\mathbf{y}_{ij}$ ($j = 1, \ldots, m_i$) known to come from $G_i$ ($i = 1, \ldots, g$), and $m = \sum_{i=1}^{i=g} m_i$. We shall refer to each $\mathbf{y}_{ij}$ as a classified observation in the sense that its population of origin is known. An example of this situation is the case study to be reported in Section 4.7 on the estimation of mixing proportions. There the classified data $\mathbf{y}_{ij}$ ($j = 1, \ldots, m_i$) have been obtained by sampling separately from $G_i$ ($i = 1, \ldots, g$), and so provide no information about the unknown mixing proportions $\pi_1, \ldots, \pi_g$, which have to be estimated solely on the basis of the unclassified data. Another example might occur in a discriminant analysis context, where the existence of the populations $G_1, \ldots, G_g$ is thus known, but where it is very difficult to obtain data of known origin. The primary aim then is not to estimate the mixing proportions, although they will have to be estimated along the way, but rather to use the unclassified data to improve the initial estimate of the component densities $f_i(\mathbf{x};\theta)$ and hence the performance of the discriminant rule, as assessed by its overall error rate in allocating a subsequent unclassified observation.

The question of whether it is a worthwhile exercise to update a discriminant rule on the basis of a limited number of unclassified observations has been considered by McLachlan and Ganesalingam (1982) for a mixture of two normal distributions. Updating procedures appropriate for nonnormal situations have been suggested by Murray and Titterington (1978) who expounded various approaches using distribution-free kernel methods, and Anderson (1979) who gave a method for the logistic discriminant function. A Bayesian approach to the problem was considered by Titterington (1976) who also considered sequential updating. The more recent work of Smith and Makov (1978) and their other papers on the Bayesian approach to the finite mixture problem, where the observations are obtained sequentially, are covered in Titterington, Smith and Makov (1985, Chapter 6).

Concerning the computation of the likelihood estimate of $\phi$ on the basis of the classified and unclassified data combined, we can easily incorporate the additional information of the known origin of the $\mathbf{y}_{ij}$ into the likelihood equation. In the case where the $\mathbf{y}_{ij}$ have been sampled from the mixture $G$, the modified versions of (1.6.1) and (1.6.2) are

$$\hat{\pi}_i = (m_i + \sum_{j=1}^{n} \hat{\tau}_{ij})/(m+n) \qquad (i = 1,\ldots,g) \tag{1.11.1}$$

and

$$\sum_{i=1}^{g} \left\{ \sum_{j=1}^{m_i} \partial f_i(\mathbf{y}_{ij};\hat{\theta})/\partial\hat{\theta} + \sum_{j=1}^{n} \hat{\tau}_{ij} \partial f_i(\mathbf{x}_j;\hat{\theta})/\partial\hat{\theta} \right\} = \mathbf{0} \tag{1.11.2}$$

However, (1.6.1) and not (1.11.1) is appropriate for the estimate of $\pi_i$, where the $\mathbf{y}_{ij}$ provide no information on the mixing proportions, for example, where they have been sampled separately from each population $G_i$ $(i = 1,\ldots,g)$, or where they have been sampled from a mixture of $G_1,\ldots,G_g$ but not in the proportions $\pi_1,\ldots,\pi_g$. The actual form of (1.11.2) for a mixture of normal distributions is displayed in Section 2.1.

Likelihood estimation is facilitated by the presence of some data of known origin with respect to each of the populations. For example, it has been pointed out for the case of normal populations with unequal covariance matrices that there are singularities in the likelihood on the edge of the parameter space. However, no singularities will occur if there are more than $p$ classified observations available from each population. Also, in the absence of data of known origin, it has been seen that there can be difficulties with the choice of suitable starting values for the EM algorithm. In the presence of classified data, an obvious choice of a starting point is the likelihood estimate of $\phi$ based solely on the classified observations $\mathbf{y}_{ij}$. In the case where the $\mathbf{y}_{ij}$ provide no information on the mixing proportions, an initial estimate of $\pi_i$ can be obtained by applying a discriminant rule formed from the $\mathbf{y}_{ij}$ to the $n$ unclassified observations $\mathbf{x}_j$, and noting the proportion of the $n$ assigned to $G_i$ $(i = 1, \ldots, g)$. This proportion might be corrected for bias before being used as a starting value for $\pi_i$. This is considered in Chapter 4, which is devoted to the problem of the estimation of mixing proportions.

## 1.12 CLASSIFICATION LIKELIHOOD APPROACH TO CLUSTERING

Another likelihood based approach to clustering is what is sometimes called the classification likelihood approach, whereby $\phi$ and $\mathbf{Z} = (\mathbf{z}_1', \ldots, \mathbf{z}_n')'$ are chosen to maximize $L_C(\phi)$, the log likelihood for the complete data specification adopted in the application of the EM algorithm to the mixture model. That is, the unobservable indicator variables $\mathbf{z}_1, \ldots, \mathbf{z}_n$ are treated as unknown parameters to be estimated along with $\phi$. Accordingly, the maximization of $L_C(\phi)$ is over the set of zero-one values of the elements of $\mathbf{z}_1, \ldots, \mathbf{z}_n$ corresponding to all possible assignments of $\mathbf{x}_1, \ldots, \mathbf{x}_n$ to $g$ groups, as well as over all admissible values of $\phi$. This procedure has been considered by several authors including Hartley and Rao (1968), John (1970), Sclove (1977), Scott and Symons (1971), and Symons (1981). Unfortunately, with this procedure, the $\mathbf{z}_j$ increase in number with the number of observations, and under such conditions the maximum likelihood estimators need not be consistent. Marriott (1975) pointed out that under the standard assumption of normal distributions with common covariance matrices, this procedure gives inconsistent estimates for the parameters involved. Bryant and Williamson (1978) extended Marriott's results and showed that the method may be expected to give asymptotically biased results quite generally; see also O'Neill (1978) and Titterington (1984b).

In principle, the maximization process for the classification likelihood approach can be carried out since it is just a matter of computing the maximum value of $L_C(\phi)$ over all possible partitions of the $n$ observations to the $g$ populations $G_1, \ldots, G_g$. In some situations, for example normal populations $G_i$ with unequal covariance matrices $\Sigma_i$, the restriction that at least $p+1$ observations belong to each $G_i$ is needed to avoid the degenerate case of infinite likelihood. Unless $n$ is quite small, however, searching over all possible partitions is prohibitive. If $\tilde{\mathbf{Z}} = (\tilde{\mathbf{z}}_1', \ldots, \tilde{\mathbf{z}}_n')'$ denotes the optimal partition, then $\tilde{z}_{ij} = 1$ or 0 according as to whether

$$\tilde{\pi}_i f_i(\mathbf{x}_j; \tilde{\theta}) > \tilde{\pi}_t f_t(\mathbf{x}_j; \tilde{\theta}), \qquad (t = 1, \ldots, g; t \neq i),$$

holds or not, where $\tilde{\boldsymbol{\theta}}$ and $\tilde{\boldsymbol{\pi}} = (\tilde{\pi}_1, \ldots, \tilde{\pi}_g)'$ are the maximum likelihood estimates of $\boldsymbol{\theta}$ and $\boldsymbol{\pi}$ for $\mathbf{x}_1, \ldots, \mathbf{x}_n$ partitioned according to $\tilde{\mathbf{z}}_1, \ldots, \tilde{\mathbf{z}}_n$ respectively. Hence a solution corresponding to a local maximum can be computed iteratively by alternating a modified version of the $E$ step but the same $M$ step as described in Section 1.6 for the application of the $EM$ algorithm under the mixture model (1.4.2). In the $E$ step at the $(k+1)$th stage of the iterative process, $z_{ij}$ is replaced not by the current estimate of the posterior probability of $\mathbf{x}_j$ belonging to $G_i$, but by 1 or 0 according as to whether

$$\pi_i^{(k)} f_i(\mathbf{x}_j; \theta^{(k)}) > \pi_t^{(k)} f_t(\mathbf{x}_j; \theta^{(k)}) \qquad (t = 1, \ldots, g; \quad t \neq i)$$

holds or not.

There are other procedures for finding the solution; for example, for normal populations, the Mahalanobis distance version of MacQueen's (1967) $k$-means procedure (really $g$-means in our notation), where $\phi$ is re-estimated after each $\mathbf{x}_j$ is assigned rather than waiting until all the observations have been allocated according to the current estimate of $\phi$.

Scott and Symons (1971) showed for the classification likelihood approach applied under the normality assumption,

$$\mathbf{x}_j \sim N(\mu_i, \Sigma) \qquad \text{in} \qquad G_i \quad (i = 1, \ldots, g), \tag{1.12.1}$$

that $\tilde{\mathbf{Z}}$ corresponds to the partition which minimizes

$$n \log |\tilde{\Sigma}| - 2 \sum_{i=1}^{g} \tilde{n}_i \log \tilde{\pi}_i,$$

or equivalently

$$n \log |\mathbf{W}| - 2 \sum_{i=1}^{g} \tilde{n}_i \log \tilde{n}_i, \tag{1.12.2}$$

where

$$\mathbf{W} = \sum_{i=1}^{g} \sum_{j=1}^{n} \tilde{z}_{ij}(\mathbf{x}_j - \tilde{\boldsymbol{\mu}}_i)(\mathbf{x}_j - \tilde{\boldsymbol{\mu}}_i)'$$

denotes the pooled within cluster covariance matrix and where

$$\tilde{\pi}_i = \tilde{n}_i/n,$$

$$\tilde{\boldsymbol{\mu}}_i = \sum_{j=1}^{n} \tilde{z}_{ij}\mathbf{x}_j/\tilde{n}_i,$$

and

$$\tilde{n}_i = \sum_{j=1}^{n} \tilde{z}_{ij}$$

for $i = 1, \ldots, g$.

If the mixing proportions $\pi_i$ are taken to be equal, or equivalently a separate sampling scheme is adopted for the data, then the $|\mathbf{W}|$ criterion is obtained, as originally suggested by Friedman and Rubin (1967). It does have the tendency to produce clusters of roughly equal size, but Symons (1981) suggested that the modified version (1.12.2) appears to go some way to overcoming this. If we further assume that the common within population covariance matrix is spherical ($\boldsymbol{\Sigma} = \sigma^2\mathbf{I}$, $\mathbf{I}$ the identity matrix), then the classification likelihood approach yields the trace $\mathbf{W}$ criterion, as proposed by Edwards and Cavalli-Sforza (1965) for partitioning the data in a hierarchical manner. By considering the effect of adding a single point to the data, Marriott (1982) investigated the properties of (1.12.2) and its corresponding form for normal but heteroscedastic populations, along with $|\mathbf{W}|$, trace $\mathbf{W}$, and three other related clustering criteria. Further references on some of these clustering criteria may be found in Seber (1984, Section 7.5), who provides a useful account of partitioning methods as part of his summary of the essentials of cluster analysis.

In the event of the availability of some data of known origin as described in Section 1.11, this information can easily be incorporated into the computation of the solution under the classification likelihood approach by setting the corresponding $z_{ij}$ equal to their known values in the iterative process. In those situations where there are sufficient data of known origin to form a reliable discriminant rule, the unclassified data can be assigned simply according to this rule without proceeding further. On the other hand, one might consider iteratively updating the rule on the basis of the classified and unclassified data combined with the latter observations partitioned according to the current version of the rule, as proposed by McLachlan (1975a, 1977). This process can be viewed as applying the classification likelihood approach from a starting point equal to the estimate of $\phi$ based solely on the classified data., For $g = 2$ normal populations with prior probabilities taken to be equal, McLachlan (1975a) showed that it leads asymptotically to an optimal partition of the unclassified data. O'Neill (1976) subsequently showed how this process can be easily modified to give an asymptotically optimal solution in the case of unknown prior probabilities.

It can be seen that the mixture likelihood approach is equivalent to the classification approach with the additional assumption that $\mathbf{z}_1, \ldots, \mathbf{z}_n$ is an (unobservable) random sample from a multinomial distribution specified by (1.6.4). With this particular prior distribution for $\mathbf{Z}$, the mixture approach can be viewed as an alternative way of specifying the model in a Bayesian formulation of the problem. For work on the Bayesian approach to cluster analysis, the reader is referred to Binder (1978, 1981), Scott and Symons (1971) and Symons (1981), in addition to the references on this approach noted in the previous section.

An account of the mixture and classification likelihood approaches has been given by McLachlan (1982a). A simulation study undertaken by Ganesalingam and McLachlan (1980a) in the case of $g = 2$ populations distributed according to (1.12.1) suggests that overall the mixture likelihood approach performs quite favorably relative to the classification approach even when mixture sampling does not apply; that is, when the $\mathbf{x}_j$ have been sampled separately from each population. The apparent slight superiority of the latter approach for data from populations in approximately equal proportions is more than offset by its inferior performance for disparate representations. Recently, Bryant and Williamson (1985) derived a central limit theorem for the bias corrected estimator of $\phi$ obtained by the classification likelihood approach under (1.12.1) for separate sampling, and compared for $p = 1$ its mean squared error with that of $\hat{\phi}$ given by the mixture likelihood approach. Other references in which the mixture approach and criteria based on the classification approach have been compared include Bayne, Beauchamp, Begovich and Kane (1980) and the monograph by Mezzich and Solomon (1980), where several cluster analysis techniques were under study. In the former, thirteen clustering methods were compared for six different parameterizations of two bivariate normal populations, while in the latter, eighteen quantitative taxonomic methods were evaluated in their application to data sets from four different fields.

Various attempts have been made for assessing the number of clusters with criteria such as trace $\mathbf{W}$ or $|\mathbf{W}|$, which we have seen can be viewed as the classification likelihood approach applied for normal populations with various assumptions about the covariance matrices. For example, Marriott (1971) on the use of $|\mathbf{W}|$ as a clustering criterion, suggested taking the value of $g$ for which $g^2|\mathbf{W}|$ is a minimum as the number of populations where, for a given $g$, the minimum of $|\mathbf{W}|$ over all partitions is implied. In the univariate normal case Engelman and Hartigan (1969) used the likelihood ratio to derive a formal test of the hypothesis of $g = 2$ groups with equal variances. The resulting statistic, denoted here by $\Lambda_1$, is the maximum of the ratio of the between group sum of squares $B$ to the within group sum of squares $W$; that is,

$$\Lambda_1 = (1 - \Lambda)/\Lambda,$$

where $\Lambda$ is the univariate version of Wilks' criterion. The maximization is over all possible $2^{n-1} - 1$ partitions of the $n$ observations into $g = 2$ groups. However, $\Lambda_1$ can be determined by examining only the $n - 1$ partitions formed by ordering the observations and dividing between two successive order statistics. They tabulated the exact value of $\Lambda_{1,\alpha}$, the $\alpha$th quantile of the null distribution of $\Lambda_1$, and gave an empirically verified approximation for $\Lambda_{1,\alpha}$,

$$\log(1 + \Lambda_{1,\alpha}) \approx -\log(1 - 2/\pi) + q_\alpha(n-2)^{-1/2} + 2.4(n-2)^{-1},$$

where $q_\alpha$ is the $\alpha$th quantile of the standard normal distribution.

Lee (1979) extended this result to the multivariate case, using essentially the union-intersection principle of test construction to derive the test statistic,

$$\Lambda_p = \max_{\gamma} \Lambda_1(\gamma),$$

where $\Lambda_1(\gamma)$ is given by $\Lambda_1$ for the projection $\gamma$ of the data onto one dimension. In testing for $g = 1$ versus $g = 2$ groups, the test based on $\Lambda_p$ is equivalent to that based on the likelihood ratio statistic. However, Lee (1979) preferred $\Lambda_p$ as, although it would appear that an infinity of linear combinations $\gamma'\mathbf{x}$ has to be considered, in practice only a rather coarse grid of linear projections need be computed to isolate the region of the optimal projection, at least for $p = 2$ or 3. It was found that a good approximation to $\Lambda_{p,\alpha}$, the $\alpha$th quantile of the null distribution of $\Lambda_p$, is given by

$$\log(1 + \Lambda_{p,\alpha}) \approx -\log(1 - 2/\pi) + q_{\alpha^*}(n-2)^{-1/2} + \{2.4 + 5.2(p-1)\}(n-2)^{-1}, \tag{1.12.3}$$

where $\alpha^* = \alpha^{1/p}$. However, (1.12.3) remains a conjecture for $p > 3$.

The univariate result of Engelman and Hartigan (1969) for testing $g = 1$ versus $g = 2$ groups was generalized by Hartigan (1978) to an alternative hypothesis of $g > 2$ groups. For a sample taken from a very general distribution, the asymptotic null distribution of $\Lambda_1$ was shown to be normal.

Other related ways for deciding upon the number of groups include the criterion based on

$$\{\operatorname{trace}\mathbf{B}/(g-1)\} \,/\, \{\operatorname{trace}\mathbf{W}/(n-g)\}, \tag{1.12.4}$$

as proposed by Caliński and Harabasz (1974). It is analogous to the $F$ statistic in the univariate case, and has been used by Edwards and Cavalli-Sforza (1965) as the basis of an $F$ test in a multivariate cluster analysis. If $C(g)$ denotes the maximum value of (1.12.4) over all partitions of the minimum spanning tree into $g$ sections, then it was suggested that the "best number" of groups can be taken to be the value of $g$ for which $C(g)$ first reaches a local maximum starting from $g = 2$, or least has a comparatively rapid increase. However, if $C(g)$ increases monotonically with $g$, then there would appear to be no group structure, whereas $C(g)$ decreasing monotonically with $g$ indicates a hierarchical structure suitable to the dichotomous grouping procedure of Edwards and Cavalli-Sforza (1965). More recently, Begovich and Kane (1982) suggested assessing the number of groups by using a procedure which incorporates measurement error into a cluster analysis through the use of computer simulations. They compared their procedure with the aforementioned methods of Marriott (1971) and Caliński and Harabasz (1974).

# 2
# Mixture Models with Normal Components

## 2.1 LIKELIHOOD ESTIMATION FOR A MIXTURE OF NORMAL DISTRIBUTIONS

In practice, finite mixture models are fitted frequently with the component densities of the mixture taken to be normal; that is, for the two-way data $\mathbf{x}_1, \ldots, \mathbf{x}_n$, it is assumed that

$$\mathbf{x}_j \sim N(\boldsymbol{\mu}_i, \boldsymbol{\Sigma}_i) \quad \text{in} \quad G_i \text{ with prob. } \pi_i \ (i = 1, \ldots, g). \tag{2.1.1}$$

Under (2.1.1), the vector $\boldsymbol{\theta}$ of parameters associated with the component densities contains the elements of the mean vectors $\boldsymbol{\mu}_1, \ldots, \boldsymbol{\mu}_g$ and the distinct elements of the covariance matrices $\boldsymbol{\Sigma}_1, \ldots, \boldsymbol{\Sigma}_g$.

In this chapter, we specialize the results presented in the previous chapter on likelihood estimation for arbitrary finite mixtures to mixtures of normal distributions. The particular case of equal covariance matrices is considered in the next section. In relation to the normality assumption (2.1.1) for the component distributions, we describe in Sections 2.5 to 2.7 the very useful method of Hawkins (1981) for testing simultaneously for normality and homoscedasticity ($\boldsymbol{\Sigma}_1 = \boldsymbol{\Sigma}_2 = \cdots = \boldsymbol{\Sigma}_g$). This test is also helpful in the detection of outliers. In the final section of this chapter a robust estimation procedure is discussed for the mixture model.

It follows from (1.6.1) and (1.6.2) that under the normality assumption (2.1.1) the likelihood estimates of the $\pi_i$, $\boldsymbol{\mu}_i$, and the $\Sigma_i$ satisfy

$$\hat{\pi}_i = \sum_{j=1}^{n} \hat{\tau}_{ij}/n, \tag{2.1.2}$$

$$\hat{\boldsymbol{\mu}}_i = \sum_{j=1}^{n} \hat{\tau}_{ij}\mathbf{x}_j/n\hat{\pi}_i, \tag{2.1.3}$$

and

$$\widehat{\Sigma}_i = \sum_{j=1}^{n} \hat{\tau}_{ij}(\mathbf{x}_j - \hat{\mu}_i)(\mathbf{x}_j - \hat{\mu}_i)'/n\hat{\pi}_i \tag{2.1.4}$$

for $i = 1, \ldots g$. In these equations, the posterior probability that $\mathbf{x}_j$ belongs to $G_i$, is given by the appropriate version of (1.4.3), namely

$$\tau_{ij} = \frac{\pi_i|\Sigma_i|^{-1/2}\exp\{-\frac{1}{2}(\mathbf{x}_j - \boldsymbol{\mu}_j)'\Sigma_i^{-1}(\mathbf{x}_j - \boldsymbol{\mu}_i)\}}{\sum_t \pi_t|\Sigma_t|^{-1/2}\exp\{-\frac{1}{2}(\mathbf{x}_j - \boldsymbol{\mu}_t)'\Sigma_t^{-1}(\mathbf{x}_j - \boldsymbol{\mu}_t)\}}. \tag{2.1.5}$$

As noted in Chapter 1, the log likelihood for a mixture of normal distributions with unequal covariance matrices is unbounded, and so the maximum likelihood estimate does not exist. Each observation $\mathbf{x}_j$ gives rise to a singularity on the edge of the parameter space. However, as remarked in Section 1.8, the nonexistence of the maximum likelihood estimate does not place a caveat on proceedings, as the essential aim of likelihood estimation is to find a sequence of roots of the likelihood equation which is consistent, and hence efficient if the usual regularity conditions hold. However, with mixture models the likelihood equation will generally have multiple roots. Thus even if it is known that there exists a sequence of roots of the likelihood equation with the desired asymptotic properties, there is the problem of identifying this sequence.

It follows from the univariate results of Kiefer (1978) and Hathaway (1985) that for mixtures of normal distributions at least in the univariate case, the sequence of roots corresponding to the largest of the local maxima for each $n$ is consistent and asymptotically normal and efficient. The last two results that the asymptotic distribution of this sequence is normal with covariance matrix equal to the inverse of the expected Fisher information matrix require the additional conditions that

$$\pi_i \neq 0 \qquad (i = 1, \ldots, g)$$

and

$$(\mu_i, \sigma_i^2) \neq (\mu_j, \sigma_j^2), \qquad (i \neq j = 1, \ldots, g),$$

on writing $\Sigma_i$ as $\sigma_i^2$ in the univariate case. In the subsequent discussion of these two results, these conditions are implicitly assumed to hold.

Kiefer (1978) verified for a mixture of univariate normal distributions in the more general case of the switching regression model that there is a sequence of roots of the likelihood equation which is consistent and asymptotically efficient and normally distributed. With probability tending to one, these roots correspond to local maxima in the interior of the parameter space; see also Peters and Walker (1978). This consistent sequence of roots is essentially unique. The reader is referred to Huzurbazar (1948) and Perlman (1983) for a precise statement of the uniqueness of a consistent sequence of roots of the likelihood equation. Kiefer's verification was for $g = 2$ components, but his result will hold for $g > 2$, as noted in Hathaway (1985) who reported some univariate results on the maximization of $L(\phi)$ over the constrained parameter space

$$\Omega_c = \{\phi \in \Omega, \sigma_i \geq c\sigma_{i+1}, i = 1, \ldots, g-1, \sigma_g \geq c\sigma_1\}, \tag{2.1.6}$$

where $\Omega$ denotes the unconstrained parameter space. For any $c \in (0,1]$, Hathaway (1985) showed that the global maximizer, $\hat{\phi}_c$ of $L(\phi)$ over $\Omega_c$ exists assuming that the sample contains at least $g+1$ distinct points, and that providing the true value of $\phi$ belongs to $\Omega_c$, then $\hat{\phi}_c$ is strongly consistent for $\phi$. This implies in the unconstrained case that it is the sequence of roots of the likelihood equation corresponding to the largest of the local maxima which is strongly consistent, and also asymptotically normal and efficient. The constraint (2.1.6) is imposed to avoid the singularities in $L(\phi)$ which occur when the mean of component, say $\mu_i$, is set equal to any observed value and $\sigma_i$ tends to zero.

For mixtures of univariate normal distributions, Hathaway (1983, 1986b) and Bezdek, Hathaway and Huggins (1985) have investigated a constrained version of the EM algorithm which incorporates the constraint (2.1.6). Their simulation results suggest that it is more robust than the ordinary version against poor choices of the starting value for $\phi$ and that it avoids convergence to a singularity of $L(\phi)$. Also, the constrained parameter space appears to have a smaller number of spurious local maximizers which, as noted by Day (1969), are generated by any small number of observations grouped sufficiently close together.

Concerning mixtures of multivariate normal distributions, it would be surprising if the above univariate result that it is the sequence of roots corresponding to the largest of the local maxima which is consistent does not carry over to $p > 1$. Redner and Walker (1984) have given the necessary regularity conditions for there to exist a consistent solution of the likelihood equation (see Section 1.8), but to the authors' knowledge, these conditions have not been verified, although they should hold as they are fairly weak. Also, Hathaway (1985) has indicated how a constrained (global) maximum likelihood formulation can be given in the multivariate case by constraining all characteristic roots of $\boldsymbol{\Sigma}_i \boldsymbol{\Sigma}_j^{-1}$ $(1 \leq i \neq j \leq g)$ to be greater than or equal to some minimum value $c > 0$ (satisfied by the true value of $\phi$).

We consider now the situation described in Section 1.11, where in addition to the unclassified data $\mathbf{x}_j$ there are $m_i$ observations $\mathbf{y}_{ij}$ $(j = 1, \ldots, m_i)$ known to come from $G_i$ $(i = 1, \ldots, g)$, and $m = \sum_{i=1}^{i=g} m_i$. It follows from (1.11.1) and (1.11.2) that the modified versions of (2.1.2) to (2.1.4) under a mixture sampling scheme with respect to $G$ for the $\mathbf{y}_{ij}$ are

$$\hat{\pi}_i = \left( m_i + \sum_{j=1}^{n} \hat{\tau}_{ij} \right) \bigg/ (m + n), \tag{2.1.7}$$

$$\hat{\boldsymbol{\mu}}_i = \left( \sum_{j=1}^{m_i} \mathbf{y}_{ij} + \sum_{j=1}^{n} \hat{\tau}_{ij} \mathbf{x}_j \right) \bigg/ \left( m_i + \sum_{j=1}^{n} \hat{\tau}_{ij} \right), \tag{2.1.8}$$

and

$$\hat{\boldsymbol{\Sigma}}_i = \frac{\{\sum_{j=1}^{m_i} (\mathbf{y}_{ij} - \hat{\boldsymbol{\mu}}_i)(\mathbf{y}_{ij} - \hat{\boldsymbol{\mu}}_i)' + \sum_{j=1}^{n} \hat{\tau}_{ij} (\mathbf{x}_j - \hat{\boldsymbol{\mu}}_i)(\mathbf{x}_j - \hat{\boldsymbol{\mu}}_i)'\}}{\left( m_i + \sum_{j=1}^{n} \hat{\tau}_{ij} \right)} \tag{2.1.9}$$

for $i = 1, \ldots, g$. In the case where the $\mathbf{y}_{ij}$ provide no information on the mixing proportions, the original equation (2.1.2) for the $\hat{\pi}_i$ should be used in conjunction with (2.1.8) and (2.1.9). As explained in Section 1.11, the presence of more than $p$ observations of known origin from each $G_i$ prevents the occurrence of singularities in the likelihood on the edge of the parameter space. The FORTRAN program given in the Appendix for likelihood estimation via the EM algorithm for a mixture of normal distributions with two-way unclassified data allows for the case where the covariance matrices are taken to be the same. There is a second program for the case where there are also available classified data from each of the possible groups.

## 2.2 NORMAL HOMOSCEDASTIC COMPONENTS

Under the normality assumption (2.1.1) with equal covariance matrices for the component distributions,

$$\mathbf{x}_j \sim N(\boldsymbol{\mu}_i, \boldsymbol{\Sigma}) \qquad \text{in} \qquad G_i \text{ with prob. } \pi_i \ (i = 1, \ldots, g), \tag{2.2.1}$$

the maximum likelihood estimate of $\boldsymbol{\phi} = (\boldsymbol{\pi}', \boldsymbol{\theta}')'$ exists and is strongly consistent; $\boldsymbol{\theta}$ denotes now the elements of the $\boldsymbol{\mu}_i$ and the distinct elements of the common covariance matrix $\boldsymbol{\Sigma}$. As remarked in Section 1.8, Redner (1981) noted that under the conditions of Wald (1949), the maximum likelihood estimator is strongly consistent for finite mixture distributions where attention is restricted to a compact subset of the parameter space. But it was pointed out to us by Perlman (1984) that the strong consistency of the maximum likelihood estimator under (2.2.1), even for the unrestricted (noncompact) parameter space, follows from Kiefer and Wolfowitz (1956). Their conditions require that the mixture density converges, though not necessarily to zero, as $\boldsymbol{\phi}$ tends to the boundary of the parameter space. To overcome the unboundedness of the density here, it is necessary to apply the device in Section 6 of Kiefer and Wolfowitz (1956), as discussed in Perlman (1972). Using this device, one works with the joint density of $g + p$ observations instead of the density corresponding to a single observation. This device was employed by Hathaway (1985) in his proof of the strong consistency of his constrained maximum likelihood estimator for univariate normal mixtures.

The choice of root of the likelihood equation is therefore straightforward under homoscedasticity, in the sense that the maximum likelihood estimator exists and is known to be consistent. The simulation results performed by Day (1969) for (2.2.1) with $g = 2$ suggest that there is no difficulty in locating the global maximum for $p = 1$ and 2, but for $p \geq 3$ there are problems with multiple maxima, particularly for small values (less than two, say) of the Mahalanobis distance between $G_1$ and $G_2$,

$$\Delta = \left\{(\boldsymbol{\mu}_1 - \boldsymbol{\mu}_2)'\boldsymbol{\Sigma}^{-1}(\boldsymbol{\mu}_1 - \boldsymbol{\mu}_2)\right\}^{1/2},$$

when $n$ is not large.

Under the normal homoscedastic model (2.2.1), the maximum likelihood estimates of $\pi_i$, $\boldsymbol{\mu}_i$, and $\boldsymbol{\Sigma}$ satisfy (2.1.2), (2.1.3), and

$$\hat{\boldsymbol{\Sigma}} = \sum_{i=1}^{g}\sum_{j=1}^{n} \hat{\tau}_{ij}(\mathbf{x}_j - \hat{\boldsymbol{\mu}}_i)(\mathbf{x}_j - \hat{\boldsymbol{\mu}}_i)'/n,$$

where in forming $\hat{\tau}_{ij}$ from (2.1.5), $\mathbf{\Sigma}_i$ is replaced by $\mathbf{\Sigma}$ $(i = 1, \ldots, g)$. In the particular case of $g = 2$, it is worth noting that Day (1969) has shown that considerable computing time can be saved by an appropriate reparameterization. To see this, first note that under (2.2.1), $\tau_{ij}$ reduces to

$$\tau_{ij} = \exp(\omega_i' \mathbf{x}_j + \omega_{0i}) \Big/ \left\{ \sum_{t=1}^{g} \exp(\omega_t' \mathbf{x}_j + \omega_{0t}) \right\}, \tag{2.2.2}$$

where

$$\omega_t = \mathbf{\Sigma}^{-1}(\boldsymbol{\mu}_t - \boldsymbol{\mu}_1)$$

and

$$\omega_{0t} = -\tfrac{1}{2}\omega_t'(\boldsymbol{\mu}_1 + \boldsymbol{\mu}_t) + \log(\pi_t/\pi_1)$$

for $t = 1, \ldots, g$; that is, $\omega_1 = \mathbf{0}$ and $\omega_{01} = 0$. Hence for $g = 2$, the likelihood can be reparameterized in terms of $\omega$, $\omega_0$, $\boldsymbol{\mu}$, and $\mathbf{V}$, where

$$\boldsymbol{\mu} = \pi_1 \boldsymbol{\mu}_1 + \pi_2 \boldsymbol{\mu}_2$$

and

$$\mathbf{V} = \mathbf{\Sigma} + \pi_1 \pi_2 (\boldsymbol{\mu}_1 - \boldsymbol{\mu}_2)(\boldsymbol{\mu}_1 - \boldsymbol{\mu}_2)'$$

are the mean and covariance matrix of the mixture distribution; $\omega$ and $\omega_0$ denote $\omega_2$ and $\omega_{02}$ with the last subscript suppressed since $g = 2$ only. The likelihood equation is now equivalent to the set

$$\hat{\boldsymbol{\mu}} = \sum_{j=1}^{n} \mathbf{x}_j / n, \tag{2.2.3}$$

$$\hat{\mathbf{V}} = \sum_{j=1}^{n} (\mathbf{x}_j - \hat{\boldsymbol{\mu}})(\mathbf{x}_j - \hat{\boldsymbol{\mu}})'/n, \tag{2.2.4}$$

$$\hat{\omega} = \hat{\mathbf{V}}^{-1}(\hat{\boldsymbol{\mu}}_2 - \hat{\boldsymbol{\mu}}_1)/\{1 - \hat{\pi}_1 \hat{\pi}_2 (\hat{\boldsymbol{\mu}}_1 - \hat{\boldsymbol{\mu}}_2)' \hat{\mathbf{V}}^{-1} (\hat{\boldsymbol{\mu}}_1 - \hat{\boldsymbol{\mu}}_2)\}, \tag{2.2.5}$$

and

$$\hat{\omega}_0 = -\tfrac{1}{2}\hat{\omega}'(\hat{\boldsymbol{\mu}}_1 + \hat{\boldsymbol{\mu}}_2) + \log(\hat{\pi}_2/\hat{\pi}_1), \tag{2.2.6}$$

where $\hat{\pi}_i$ and $\hat{\boldsymbol{\mu}}_i$ are given by (2.1.2) and (2.1.3) respectively with $\hat{\tau}_{ij}$ formed from (2.2.2) for $i = 1, 2$. As $\hat{\boldsymbol{\mu}}$ and $\hat{\mathbf{V}}$ are given explicitly, it can be seen from (2.2.3) to (2.2.6) that only the values of $\widehat{\boldsymbol{\omega}}$ and $\hat{\omega}_0$ have to be recomputed on each $M$ step of the EM algorithm.

Ganesalingam and McLachlan (1981) considered the case where, in addition to the unclassified data, there are some classified observations $\mathbf{y}_{ij}$ ($j = 1, \ldots, m_i$) available from $G_i$ ($i = 1, 2$), and $m = m_1 + m_2$. The modified equations are

$$\hat{\boldsymbol{\mu}} = \left( \sum_{i=1}^{2} \sum_{j=1}^{m_i} \mathbf{y}_{ij} + \sum_{j=1}^{n} \mathbf{x}_j \right) \Bigg/ (m+n),$$

$$\hat{\mathbf{V}} = \left\{ \sum_{i=1}^{2} \sum_{j=1}^{m_i} (\mathbf{y}_{ij} - \hat{\boldsymbol{\mu}})(\mathbf{y}_{ij} - \hat{\boldsymbol{\mu}})' + \sum_{j=1}^{n} (\mathbf{x}_j - \hat{\boldsymbol{\mu}})(\mathbf{x}_j - \hat{\boldsymbol{\mu}})' \right\} \Bigg/ (m+n),$$

$$\widehat{\boldsymbol{\omega}} = \hat{\mathbf{V}}^{-1}(\hat{\boldsymbol{\mu}}_2 - \hat{\boldsymbol{\mu}}_1)/\{1 - \hat{\pi}_1^+ \hat{\pi}_2^+ (\hat{\boldsymbol{\mu}}_1 - \hat{\boldsymbol{\mu}}_2)' \hat{\mathbf{V}}^{-1} (\hat{\boldsymbol{\mu}}_1 - \hat{\boldsymbol{\mu}}_2)\},$$

and

$$\hat{\omega}_0 = -\tfrac{1}{2} \widehat{\boldsymbol{\omega}}' (\hat{\boldsymbol{\mu}}_1 + \hat{\boldsymbol{\mu}}_2) + \log(\hat{\pi}_2 / \hat{\pi}_1),$$

where $\hat{\pi}_i^+$ and $\hat{\boldsymbol{\mu}}_i$ are given by (2.1.7) and (2.1.8) respectively ($i = 1, 2$). If the classified data were sampled from the mixture $G$, then $\hat{\pi}_i = \hat{\pi}_i^+$, but if they provide no information on the mixing proportions, then $\hat{\pi}_i$ is given by (2.1.2) for $i = 1, 2$.

## 2.3 ASYMPTOTIC RELATIVE EFFICIENCY OF THE MIXTURE LIKELIHOOD APPROACH

We consider here the relative efficiency of the mixture likelihood approach to forming an estimate of the optimal discriminant rule from unclassified data in the situation where the superpopulation $G$ actually consists of $g$ distinct populations $G_1, \ldots, G_g$, as discussed in Section 1.11. In this context it is assumed that the estimates produced by the mixture likelihood approach are correctly identified with the externally existing populations $G_1, \ldots, G_g$. This last matter is to be discussed further in Section 5.1 on the problem of assessing the performance of the mixture likelihood approach to clustering. As noted in Section 1.4, the optimal discriminant rule $R(\mathbf{x}; \boldsymbol{\phi})$ assigns an unclassified observation $\mathbf{x}$ to the population to which it has the highest posterior probability of belonging.

Results for the asymptotic relative efficiency of the mixture likelihood approach are available for a mixture of two normal distributions with equal covariance matrices. In fitting a mixture model to the data $\mathbf{X} = (\mathbf{x}_1', \ldots, \mathbf{x}_n')'$ under the normal homoscedastic model (2.2.1) with $g = 2$, the rule $R(\mathbf{x}; \hat{\phi})$ which can be formed after computation of the likelihood estimate $\hat{\phi}$, reduces to assigning $\mathbf{x}$ to either $G_1$ or $G_2$ according as to whether

$$\hat{\omega}'\mathbf{x} + \hat{\omega}_0 \tag{2.3.1}$$

is less or greater than zero, where $\hat{\omega}$ and $\hat{\omega}_0$ are defined as in the previous section. The efficiency of the mixture likelihood approach is to be defined relative to the version of (2.3.1) appropriate where the data $\mathbf{x}_j$ are of known origin with respect to $G_1, \ldots, G_g$ respectively; that is, where the associated indicator vectors in $\mathbf{Z} = (\mathbf{z}_1', \ldots, \mathbf{z}_n')'$ are known. Let $\hat{\phi}_C$ be the likelihood estimate of $\phi$ computed from $L_C(\phi)$, the log likelihood for the complete data $\mathbf{X}$ and $\mathbf{Z}$, as given by (1.6.5) in the formulation of the application of the EM algorithm to the mixture problem in Section 1.6. The associated rule $R(\mathbf{x}; \hat{\phi}_C)$ is then based on the corresponding version of (2.3.1), which is the familiar linear discriminant function

$$\hat{\omega}_C'\mathbf{x} + \hat{\omega}_{0C}, \tag{2.3.2}$$

where

$$\hat{\omega}_C = \{(n-2)\mathbf{S}/n\}^{-1}(\bar{\mathbf{x}}_2 - \bar{\mathbf{x}}_1) \tag{2.3.3}$$

and

$$\hat{\omega}_{0C} = -\tfrac{1}{2}\hat{\omega}_C'(\bar{\mathbf{x}}_1 + \bar{\mathbf{x}}_2) + \log(n_2/n_1), \tag{2.3.4}$$

and where

$$\bar{\mathbf{x}}_i = \sum_{j=1}^{n} z_{ij}\mathbf{x}_j/n_i \qquad (i = 1, 2),$$

$$\mathbf{S} = \sum_{i=1}^{2}\sum_{j=1}^{n} z_{ij}(\mathbf{x}_j - \bar{\mathbf{x}}_i)(\mathbf{x}_j - \bar{\mathbf{x}}_i)'/(n-2),$$

and

$$n_i = \sum_{j=1}^{n} z_{ij} \qquad (i = 1, 2).$$

Apart from the divisor $n$ and not $(n-2)$ in the estimate of $\Sigma$ in (2.3.3) and the term $\log(n_2/n_1)$ in (2.3.4), (2.3.2) is Fisher's linear discriminant function as modified by Anderson (1951).

Concerning the application of $R(\mathbf{x}; \hat{\phi})$ to a randomly chosen observation $\mathbf{x}$ from the mixture $G$, where $\mathbf{x}$ is distributed independently of the original data $\mathbf{X}$, its overall misallocation rate conditional on $\mathbf{X}$, or equivalently $\hat{\phi}$, is given by

$$e(\phi; \hat{\phi}) = \sum_{i=1}^{2} \pi_i \Phi \left\{(-1)^{i+1}(\hat{\omega}'\boldsymbol{\mu}_i + \hat{\omega}_0)/(\hat{\omega}'\Sigma\hat{\omega})^{1/2}\right\},$$

where $\Phi$ denotes the standard normal distribution function. Similarly, $e(\phi; \hat{\phi}_C)$ denotes the overall conditional misallocation rate for $R(\mathbf{x}; \hat{\phi}_C)$. The overall unconditional misallocation rates, $E\{e(\phi; \hat{\phi})\}$ and $E\{e(\phi; \hat{\phi}_C)\}$, are defined by taking the expectation of $e(\phi; \hat{\phi})$ and of $e(\phi; \hat{\phi}_C)$ over the sampling distribution of $\hat{\phi}$ and of $\hat{\phi}_C$ respectively.

For a mixture of two normal distributions, Ganesalingam and McLachlan (1978) defined the asymptotic relative efficiency of the mixture approach by considering the ratio

$$\varepsilon = [E\{e(\phi; \hat{\phi}_C)\} - e_0]/[E\{e(\phi; \hat{\phi})\} - e_0], \tag{2.3.5}$$

where $e_0 = e(\phi; \phi)$ is the common limiting value as $n \to \infty$ of the unconditional misallocation rates. The asymptotic relative efficiency was obtained in the univariate case by evaluating the numerator and denominator of (2.3.5) up to and including terms of order $1/n$. The multivariate analogue of this problem was investigated independently by O'Neill (1978). By its definition, the asymptotic relative efficiency of the mixture likelihood approach does not depend on $n$, and O'Neill (1978) showed that it also does not depend on the dimensions $p$ for equal mixing proportions $(\pi_1 = \pi_2 = 0.5)$.

For selected combinations of the parameters $\pi_1$, $p$, $n$, and the Mahalanobis distance $\Delta$ between $G_1$ and $G_2$, the asymptotic relative efficiency is listed as a percentage in Table 1, along with its simulated value as taken from Ganesalingam and McLachlan (1979a). A comparison of the asymptotic versus simulated values of $\varepsilon$ in Table 1 indicate that the asymptotic relative efficiency does not give a reliable guide as to the true performance of $R(\mathbf{x}; \hat{\phi})$ relative to the version $R(\mathbf{x}; \hat{\phi}_C)$ formed from classified data when $n$ is small, particularly for $\Delta = 1$ representing populations close together. This is not surprising, as it is well known (Day, 1969 and Hosmer, 1973a) that $n$ has to be very large for the mixture problem before the asymptotic likelihood theory is applicable. Further simulation studies by Ganesalingam and McLachlan (1979a) in the univariate case suggest that the asymptotic relative efficiency gives reliable predictions for $p = 1$ at least for $n \geq 100$ and for populations at least moderately separated ($\Delta \geq 2$, say). It would appear from the simulated values of $\varepsilon$ in Table 1 that in order for $R(\mathbf{x}; \hat{\phi})$ to have an overall allocation rate comparable to that of $R(\mathbf{x}; \hat{\phi}_C)$, $\hat{\phi}$ needs to be formed from unclassified data with about two to five times as many observations as in the computation of $\hat{\phi}_C$ from a classified set of data.

More recently, Amoh (1985) investigated the relative efficiency of the discriminant rule estimated from unclassified data arising from a mixture of two inverse Gaussian distributions.

## 2.4 EXPECTED AND OBSERVED INFORMATION MATRICES

Hill (1963) gave a general power series expansion of the expected Fisher information about the mixing proportion $\pi_1$ in a univariate mixture of two known densities, normal with equal variances. Behboodian (1972) presented a method for the numerical calculation of the expected information matrix $\mathcal{I}(\phi)$ for a mixture of univariate normal distributions with arbitrary variances, while Tan and Chang (1972) considered the homoscedastic case in deriving the asymptotic relative efficiency of the moment estimator of

**TABLE 1** Asymptotic versus simulation results in parentheses for the relative efficiency of the mixture likelihood approach

| | $p = 1, n = 20$ | | $p = 2, n = 20$ | | $p = 3, n = 40$ | |
|---|---|---|---|---|---|---|
| $\Delta$ | $\pi_1 = 0.25$ | $\pi_1 = 0.50$ | $\pi_1 = 0.25$ | $\pi_1 = 0.50$ | $\pi_1 = 0.25$ | $\pi_1 = 0.50$ |
| 1 | 0.25 | 0.51 | 0.34 | 0.51 | 0.42 | 0.51 |
| | (33.01) | (25.12) | (46.71) | (63.11) | (25.00) | (43.39) |
| 2 | 7.29 | 10.08 | 9.36 | 10.08 | 10.51 | 10.08 |
| | (22.05) | (17.74) | (25.73) | (16.26) | (16.28) | (14.51) |
| 3 | 31.41 | 35.92 | 35.13 | 35.92 | 36.78 | 35.92 |
| | (19.57) | (23.54) | (43.91) | (29.63) | (29.01) | (23.46) |

Source: Ganesalingam and McLachlan (1979a).

$\phi$. For two univariate normal populations with no restrictions on the variances, Hosmer and Dick (1977) studied the expected information matrix for $\phi$, where in addition to the unclassified observations $x_1, \ldots, x_n$, there were also some observations of known origin obtained under various sampling schemes.

In the multivariate case for a mixture of two normal distributions with a common covariance matrix, Chang (1976, 1979) showed how the evaluation of $\mathcal{I}(\phi)$ can be simplified to various calculations, each equivalent to finding the information matrix for a three-dimensional mixture. For the same mixture but where in addition there are also some data of known origin sampled from the mixture, O'Neill (1978) obtained the expected information matrix for the discriminant function coefficients in $\omega$ and $\omega_0$. For a similar model but where the classified data were sampled separately from each of the two populations, Ganesalingam and McLachlan (1981) considered the expected information matrix for $\phi$ in deriving the asymptotic variance of the mixing proportion estimator $\hat{\pi}_1$.

The main purpose in calculating the expected Fisher information matrix for $\phi$ in the above papers on two-component normal mixtures was to obtain the asymptotic covariance matrix of $\hat{\phi}$ either to study the performance of $\hat{\phi}$ directly or to compute the asymptotic relative efficiencies of other estimators of $\phi$. In practice $\mathcal{I}(\hat{\phi})$ will provide an estimate of the covariance matrix of $\hat{\phi}$, but it is much easier to compute the observed information matrix $\mathbf{I}(\hat{\phi})$, where

$$\mathbf{I}(\phi) = ((-\partial^2 L(\phi)/\partial\phi_i\partial\phi_j)).$$

As noted in Section 1.9, $\mathbf{I}(\hat{\phi})$ can be approximated directly in terms of the gradient vector of the log likelihood function $L_C(\phi)$ for the complete data problem. It can be seen from (1.9.3), that for the unclassified data $\mathbf{x}_1, \ldots, \mathbf{x}_n$, $\mathbf{I}(\hat{\phi})$ is approximated as

$$\mathbf{I}(\hat{\phi}) \approx \sum_{j=1}^{n} \hat{\mathbf{h}}_j \hat{\mathbf{h}}_j', \tag{2.4.1}$$

where

$$\hat{\mathbf{h}}_j = \partial L_C(\hat{\phi}; \mathbf{x}_j, \hat{\boldsymbol{\tau}}_j)/\partial\hat{\phi}. \tag{2.4.2}$$

We consider now the computation of $\hat{\mathbf{h}}_j$, the gradient vector at $\phi = \hat{\phi}$ of the log likelihood for the single data point $\mathbf{x}_j$ and $\mathbf{z}_j$, where the unknown indicator vector $\mathbf{z}_j$ is replaced by the vector of the estimated posterior probabilities of group membership, $\hat{\boldsymbol{\tau}}_j = (\hat{\tau}_{1j}, \ldots, \hat{\tau}_{gj})'$ for $\mathbf{x}_j$ $(j = 1, \ldots, n)$.

Let $\boldsymbol{\pi}_{(g)} = (\pi_1, \ldots, \pi_{g-1})'$ and $\boldsymbol{\xi}_i$ be the vector containing the $\frac{1}{2}p(p+1)$ distinct elements of $\boldsymbol{\Sigma}_i$ $(i = 1, \ldots, g)$. Then partition $\phi$ as

$$\phi = (\boldsymbol{\pi}'_{(g)}, \boldsymbol{\mu}'_1, \ldots, \boldsymbol{\mu}'_g, \boldsymbol{\xi}'_1, \ldots, \boldsymbol{\xi}'_g)',$$

and let

$$\hat{\mathbf{h}}_j = (\hat{\mathbf{h}}'_{\pi j}, \hat{\mathbf{h}}'_{\mu 1j}, \ldots, \hat{\mathbf{h}}'_{\mu gj}, \hat{\mathbf{h}}'_{\xi 1j}, \ldots, \hat{\mathbf{h}}'_{\xi gj})'$$

be the corresponding partition of $\hat{\mathbf{h}}_j$ $(j = 1, \ldots, n)$. It follows on differentiation that the $k$th element of $\hat{\mathbf{h}}_{\pi j}$ is given by

$$(\hat{\mathbf{h}}_{\pi j})_k = \hat{\tau}_{kj}/\hat{\pi}_1 - \hat{\tau}_{gj}/\hat{\pi}_g \qquad (k = 1, \ldots, g-1)$$

and

$$\hat{\mathbf{h}}_{\mu ij} = \hat{\tau}_{ij}\hat{\boldsymbol{\Sigma}}_i^{-1}(\mathbf{x}_j - \hat{\boldsymbol{\mu}}_i) \qquad (i = 1, \ldots, g)$$

for $j = 1, \ldots, n$.

Considering the computation of $\hat{\mathbf{h}}_{\xi ij}$, let $\delta_{rs}$ be the Kronecker delta, and write

$$\boldsymbol{\Sigma}_i^{-1} = (\boldsymbol{\sigma}_i^{(1)}, \ldots, \boldsymbol{\sigma}_i^{(p)})$$

for $i = 1, \ldots, g$, where $\boldsymbol{\sigma}_i^{(r)}$ denotes the $r$th column of $\boldsymbol{\Sigma}_i^{-1}$. If the $k$th element of $\hat{\mathbf{h}}_{\xi ij}$ corresponds to differentiation with respect to, say $(\boldsymbol{\Sigma}_i)_{rs}$, the $(r,s)$th element of $\boldsymbol{\Sigma}_i$ $(r \leq s)$, then

$$(\hat{\mathbf{h}}_{\xi ij})_k = \tfrac{1}{2}\hat{\tau}_{ij}(2 - \delta_{rs}) \times \left[-\left(\hat{\boldsymbol{\Sigma}}_i^{-1}\right)_{rs} + \left\{(\mathbf{x}_j - \hat{\boldsymbol{\mu}}_i)'\hat{\boldsymbol{\sigma}}_i^{(r)}\right\}\left\{(\mathbf{x}_j - \hat{\boldsymbol{\mu}}_i)'\hat{\boldsymbol{\sigma}}_i^{(s)}\right\}\right]$$

for $i = 1, \ldots, g$ and $j = 1, \ldots, n$.

Suppose that in addition to the unclassified data there are some $m$ observations of known origin with respect to the component populations $G_1, \ldots, G_g$. For this purpose we shall let these classified data be denoted by $\mathbf{y}_1, \ldots, \mathbf{y}_m$ with $\mathbf{z}_{y_1}, \ldots, \mathbf{z}_{y_m}$ the known associated vectors of indicator variables. Then the total observed information matrix for the unclassified and classified data combined is equal to

$$\mathbf{I}(\hat{\phi}) + \mathbf{I}_C(\hat{\phi}),$$

where

$$\mathbf{I}_C(\phi) = ((-\partial^2 L_C(\phi)/\partial\phi_i\partial\phi_j)).$$

The observed information matrix $\mathbf{I}(\hat{\phi})$ for the unclassified data can be computed as above, and it is computationally attractive to compute $\mathbf{I}_C(\hat{\phi})$ using the approximation

$$\mathbf{I}_C(\hat{\phi}) \approx \sum_{j=1}^{m} \tilde{\mathbf{h}}_j \tilde{\mathbf{h}}_j' \qquad (2.4.3)$$

where

$$\tilde{\mathbf{h}}_j = \partial L_C(\hat{\phi}; \mathbf{y}_j, \mathbf{z}_{y_j})/\partial\hat{\phi} \qquad (j = 1, \ldots, m).$$

That is, for a classified observation, we proceed in the same manner as if it were unclassified except that we use its known indicator vector in (2.4.2) instead of an estimate given by the vector of the estimated posterior probabilities.

## 2.5 ASSESSMENT OF NORMALITY FOR COMPONENT DISTRIBUTIONS: PARTIALLY CLASSIFIED DATA

Up to now, attention has been focused on likelihood estimation for the mixture model under the parametric forms specified for the component distributions. An important consideration in practice is the applicability of the adopted forms for the component densities. A common assumption is to take the component densities to be normal, as adopted in the previous sections of this chapter.

We consider first the problem of assessing normality for the component distributions of a mixture in the now straightforward situation where, in addition to the unclassified data $\mathbf{x}_1, \ldots, \mathbf{x}_n$, there are $m = \sum_{i=1}^{i=g} m_i$ observations $\mathbf{y}_{ij}$ $(j = 1, \ldots, m_i)$ known to come from $G_i$ $(i = 1, \ldots, g)$. The more difficult problem where there are no classified data, as in a typical cluster anaylsis example, will be discussed later in Section 2.6. An obvious way of proceeding in the former case is to assess the normality assumption for each individual population $G_i$ ($i = 1, \ldots, g$), as in a discriminant analysis context. Once this matter has been satisfactorily resolved, the assumption that the unclassified data $\mathbf{x}_1, \ldots, \mathbf{x}_n$ come from a mixture of the $G_i$ can be approached by introducing a measure of typicality to assess how typical each $\mathbf{x}_j$ is with respect to each $G_i$. This last process will be described in the next section. It is assumed here that each $m_i$ is sufficiently large to provide reasonable estimates of the unknown parameters in $f_i(\mathbf{x}; \boldsymbol{\theta})$, $i = 1, \ldots, g$. If some or all of the $m_i$ are very small, then the unclassified data will have to be used too in forming these estimates, and so the assessment procedure therefore is essentially the same as for an unclassified sample.

An extensive review of both formal and informal methods for testing for normality has been given by Gnanadesikan (1977, Section 5.4.2), Cox and Small (1978), and Mardia (1980). More recently, the problem has been considered by Bera and John (1983), Hall and Welsh (1983), Koziol (1982, 1983, 1986), Machado (1983), Matthews (1984), Royston (1983), Small (1980), and Srivastava (1984), among others. In the present discriminant analysis framework where more than one population is under assessment, there is also the question of whether the populations have a common covariance matrix (homoscedasticity) to be considered. A convenient test therefore in this framework is the comparatively new procedure proposed by Hawkins (1981) which can be used to test simultaneously for normality and homoscedasticity; see also Fatti, Hawkins and Raath (1982). Besides being easily implemented, this test has comparable power with standard inferential procedures such as the likelihood ratio test statistic for homoscedasticity used in conjunction with the multivariate coefficients of skewness and kurtosis (Mardia, 1974) for normality.

As we shall be using Hawkins' test to assess model fit for some of the data sets to be analysed in later sections, we now give a brief description of this technique for testing the hypothesis

$$H_0 : \mathbf{x} \sim N(\boldsymbol{\mu}_i, \boldsymbol{\Sigma}) \quad \text{in} \quad G_i \ (i = 1, \ldots, g), \tag{2.5.1}$$

on the basis of the classified data $\mathbf{y}_{ij}$ ($i = 1, \ldots, g; j = 1, \ldots, m_i$). Let

$$\bar{\mathbf{y}}_i = \sum_{j=1}^{m_i} \mathbf{y}_{ij} / m_i \tag{2.5.2}$$

and

$$\mathbf{S}_i = \sum_{j=1}^{m_i} (\mathbf{y}_{ij} - \bar{\mathbf{y}}_i)(\mathbf{y}_{ij} - \bar{\mathbf{y}}_i)'/(m_i - 1) \tag{2.5.3}$$

for $i = 1, \ldots, g$, and let

$$\mathbf{S} = \sum_{i=1}^{g} (m_i - 1)\mathbf{S}_i/(m - g).$$

The Mahalanobis squared distance between $\mathbf{y}_{ij}$ and $\bar{\mathbf{y}}_i$ with respect to $\mathbf{S}$ is denoted by $D(\mathbf{y}_{ij}, \bar{\mathbf{y}}_i; \mathbf{S})$, so that

$$D(\mathbf{y}_{ij}, \bar{\mathbf{y}}_i; \mathbf{S}) = (\mathbf{y}_{ij} - \bar{\mathbf{y}}_i)'\mathbf{S}^{-1}(\mathbf{y}_{ij} - \bar{\mathbf{y}}_i).$$

This notation for the Mahalanobis squared distance between two vectors with respect to some positive definite symmetric matrix is to be used throughout this book. For a given population $G_i$, the test considers the Mahalanobis squared distance between each $\mathbf{y}_{ij}$ $(j = 1, \ldots, m_i)$ and the mean of the sample from $G_i$, but where each $\mathbf{y}_{ij}$ is deleted from the sample in case it severely contaminates the estimates of the mean and covariance matrix of $G_i$ $(i = 1, \ldots, g)$. Accordingly, the Mahalanobis squared distance

$$D(\mathbf{y}_{ij}, \bar{\mathbf{y}}_{i(ij)}; \mathbf{S}_{(ij)}) \tag{2.5.4}$$

is computed, where $\bar{\mathbf{y}}_{i(ij)}$ and $\mathbf{S}_{(ij)}$ denote the resulting values of $\bar{\mathbf{y}}_i$ and $\mathbf{S}$ after the deletion of $\mathbf{y}_{ij}$ from the data. Under $H_0$, it follows that

$$c(m_i, \nu) D(\mathbf{y}_{ij}, \bar{\mathbf{y}}_{i(ij)}; \mathbf{S}_{(ij)}) \tag{2.5.5}$$

is distributed according to an $F$ distribution with $p$ and $\nu = m - g - p$ degrees of freedom, where

$$c(m_i, \nu) = \{(m_i - 1)\nu\} \,/\, \{(m_i p)(\nu + p - 1)\}. \tag{2.5.6}$$

To avoid the recomputation of $\bar{\mathbf{y}}_i$ and $\mathbf{S}$ after the deletion of each $\mathbf{y}_{ij}$ from the data, it was shown that (2.5.5) can be computed using the result that

$$c(m_i,\nu)D(\mathbf{y}_{ij},\bar{\mathbf{y}}_{i(ij)};\mathbf{S}_{(ij)}) = \frac{(\nu m_i/p)D(\mathbf{y}_{ij},\bar{\mathbf{y}}_i;\mathbf{S})}{(\nu+p)(m_i-1)-m_iD(\mathbf{y}_{ij},\bar{\mathbf{y}}_i;\mathbf{S})}. \tag{2.5.7}$$

If $a_{ij}$ denotes the area to the right of the observed value of (2.5.7) under the $F_{p,\nu}$ distribution, then under $H_0$ we have that

$$H_{0i}: a_{i1},\ldots,a_{im_i} \overset{iid}{\sim} U(0,1) \qquad (i=1,\ldots,g) \tag{2.5.8}$$

holds approximately, where $U(0,1)$ denotes the uniform distribution on the unit interval. The result (2.5.8) is only approximate as for a given $i$, the $a_{ij}$ are only independent exactly as $m_i \to \infty$, due to the presence of the estimates of $\boldsymbol{\mu}_i$ and $\boldsymbol{\Sigma}$ in the formation of (2.5.4). Hawkins (1981) has reported empirical evidence which suggests that subsequent steps in his test which treat (2.5.8) as if it were an exact result should be approximately valid. He also noted that (2.5.8) can be valid for nonnormal component distributions but it is likely that such cases would be very rare.

A close inspection of the tail areas $a_{ij}$ including $Q$-$Q$ plots can be undertaken to detect departures from the $g$ hypotheses $H_{0i}$, and hence from the original hypothesis $H_0$. In conjunction with this detailed analysis, Hawkins (1981) advocated the use of the Anderson-Darling statistic for assessing (2.5.8), as this statistic is particularly sensitive to fit in the tails of the distribution. It can be computed for the sample of $m_i$ values $a_{ij}$ $(j=1,\ldots,m_i)$ by

$$W_i = -m_i - \sum_{j=1}^{m_i}(2j-1)\left\{\log a_{i(j)} + \log(1-a_{i(m_i-j+1)})\right\} \Big/ m_i$$

$$(i=1,\ldots,g),$$

where for each $i$, $a_{i(1)} \le a_{i(2)} \le \cdots \le a_{i(m_i)}$ denote the $m_i$ order statistics of the $a_{ij}$. In the asymptotic resolution of each $W_i$ into standard normal variates $W_{ik}$ according to

$$W_i = \sum_{k=1}^{\infty} W_{ik}^2/\{k(k+1)\} \qquad (i=1,\ldots,g),$$

attention is focused on the first two components

$$W_{i1} = -(3/m_i)^{1/2} \sum_{j=1}^{m_i} (2a_{i(j)} - 1)$$

and

$$W_{i2} = -(5/m_i)^{1/2} \sum_{j=1}^{m_i} \tfrac{1}{2}\{3(2a_{i(j)} - 1)^2 - 1\}.$$

Similarly, the Anderson-Darling statistic $W_T$ and its first two components $W_{T1}$ and $W_{T2}$ can be computed for the single sample where all the $a_{ij}$ are combined.

Some simulations performed by Hawkins (1981) suggest that the size of the test will be approximately 0.1 if (2.5.8) is rejected if any $W_i$ exceeds 2.5 (the asymptotic 95th percentile) or any $W_{ik}$ $(k = 1,2)$ exceeds 2.54 in magnitude. However, in view of (2.5.8) not holding exactly for finite $m_i$, caution should be exercised in any formal test based on the Anderson-Darling statistics $W_i$ and $W_T$ and their components.

Rather, their primary role as proposed by Hawkins (1981) is to provide quick summary statistics for interpreting qualitatively departures from $H_0$, which may manifest themselves in a number of ways in the $a_{ij}$. Nonnormality in the form of heavy tails in the data leads to a $U$ shaped distribution with an excess of the $a_{ij}$ near 0 or near 1 for a given $i$. The second component $W_{i2}$, which is essentially the variance of the $a_{ij}$ $(j = 1, \ldots, m_i)$, is useful in detecting this (and other departures from normality). The presence of heteroscedasticity generally causes there to be an excess of the $a_{ij}$ near 1 in one population, say $G_r$, with an excess near 0 for another population, say $G_t$. Depending on the degree of heteroscedasticity, $W_{r1}$ and $W_{t1}$ will be large in magnitude, but of opposite sign, reflecting the clustering of the $a_{rj}$ and $a_{tj}$ at different ends of the unit interval. For nonnormal data, $W_{r1}$ and $W_{t1}$ may not necessarily have different signs but there should be large differences between them to indicate heteroscedasticity. If the data $\mathbf{y}_{ij}$ are really normal, then the imbalances in the $a_{rj}$ and the $a_{tj}$ tend to cancel each other out in the formation of $W_T$ and its components from the combined areas $a_{ij}$. Thus significant values of $W_T$ and its components can be taken as a fair indication of nonnormality.

If this assessment finds that a normal mixture model does not adequately fit the data, then one option would be to seek a transformation of the data in an attempt to achieve normality for the component distributions. An account of this problem may be found in Gnanadesikan (1977,

Section 5.3). One approach discussed there is the generalization of the univariate technique of Box and Cox (1964), whereby for a given component $G_i$ $(i = 1, \ldots, g)$, each coordinate of $\mathbf{y}_{ij}$, $y_{ijv}$ $(v = 1, \ldots, p)$, is transformed into

$$y_{ijv}^{(\varsigma_{iv})} = \begin{cases} (y_{ijv}^{\varsigma_{iv}} - 1)/\varsigma_{iv}, & \varsigma_{iv} \neq 0, \\ \log y_{ijv}, & \varsigma_{iv} = 0. \end{cases}$$

The unknown $\varsigma_{i1}, \ldots, \varsigma_{ip}$ can be estimated by maximum likelihood on the basis of the transformed values of $\mathbf{y}_{ij}$ $(j = 1, \ldots, m_i)$ under the assumption of multivariate normality. This approach also provides another assessment of normality as the likelihood ratio test can be used to test the null hypothesis that $\varsigma_{iv} = 1$ $(v = 1, \ldots, p)$. The above transformation along with more robust methods are discussed by Lesaffre (1983), who surveyed normality tests and transformations for use in particular in discriminant analysis.

If it is concluded that it is reasonable to take the component populations $G_i$ to be normal but heteroscedastic, that is

$$\mathbf{x} \sim N(\boldsymbol{\mu}_i, \boldsymbol{\Sigma}_i) \quad \text{in} \quad G_i \quad (i = 1, \ldots, g), \tag{2.5.9}$$

we may wish to recompute the $a_{ij}$ under (2.5.9). We may still be interested in the $a_{ij}$ to assess whether there are any outliers among the $\mathbf{y}_{ij}$. This is discussed in the next section. Under (2.5.9), we modify the Mahalanobis squared distance (2.5.4) by replacing the pooled sample covariance matrix $\mathbf{S}_{(ij)}$ by $\mathbf{S}_{i(ij)}$, as computed according to (2.5.3) but with $\mathbf{y}_{ij}$ deleted. Consequently, $a_{ij}$ is defined now to be the area to the right of

$$c(m_i, \nu_i) D(\mathbf{y}_{ij}, \bar{\mathbf{y}}_{i(ij)}; \mathbf{S}_{i(ij)}) \tag{2.5.10}$$

under the $F_{p,\nu_i}$ distribution, where $\nu_i = m_i - p - 1$. The value of (2.5.10) can be computed from (2.5.7) where now $\nu$ is replaced by $\nu_i$ and $\mathbf{S}$ by $\mathbf{S}_i$.

## 2.6 ASSESSMENT OF TYPICALITY: PARTIALLY CLASSIFIED DATA

Attention is now focused on the detection of apparent outliers among the classified observations $\mathbf{y}_{ij}$ $(i = 1, \ldots, g; j = 1, \ldots, m_i)$, and the unclassified data $\mathbf{x}_j$ $(j = 1, \ldots, n)$. Generally an identification of observations which appear to stand apart from the bulk of the data is undertaken with a view to taking some action to reduce their effect in the formation of any estimates. However, in some instances, the atypical observations may be of much interest in their own right. Considering first the classified data, we have for each $\mathbf{y}_{ij}$ from the assessment of normality and homoscedasticity described in the previous section, the associated tail area $a_{ij}$ computed to the right of either (2.5.7) or (2.5.10) under the $F$ distribution with the appropriate degrees of freedom, depending on whether the normal populations are taken to have the same or different covariance matrices. If $a_{ij}$ is close to zero, then $\mathbf{y}_{ij}$ is regarded as atypical of population $G_i$. Note that under the restriction to tests invariant under arbitrary full-rank linear transformations (Hawkins, 1980, page 107), (2.5.10) is equivalent to the optimal test statistic for a single outlier $\mathbf{y}^\dagger$ for both the alternative hypotheses

$$H_{1i} : \mathbf{y}^\dagger \sim N(\mu_i^\dagger, \Sigma_i)$$

and

$$H_{2i} : \mathbf{y}^\dagger \sim N(\boldsymbol{\mu}_i, \kappa_i \Sigma_i),$$

where $\boldsymbol{\mu}_i^\dagger$ is some vector not equal to $\boldsymbol{\mu}_i$ and $\kappa_i$ is some positive constant different from unity; similarly, for the statistic (2.5.7) under homoscedasticity.

As stressed by Barnett (1983) in the discussion of a review of outliers by Beckman and Cook (1983), a more vital question than whether an observation is atypical is whether it is statistically unreasonable when assessed as an extreme (a discordant outlier); for example, in the present context with $p = 1$, whether say the largest of the $m_i$ observations from $G_i$ is unreasonably large as an observation on the $m_i$th order statistic for a sample of size $m_i$ from $G_i$. However, as acknowledged by Barnett (1983), generalizations to $p > 1$ are limited as the notion of order is ambiguous and ill-defined in multivariate samples.

One way of approaching the assessment of whether the unclassified data $\mathbf{x}_1, \ldots, \mathbf{x}_n$ are from a mixture of $G_1, \ldots, G_g$ is to assess how typical $\mathbf{x}_j$ is of each $G_i$ taken in turn ($j = 1, \ldots, n$). An observation which is atypical of each $G_i$ may well be considered a contaminant which, in the outlier terminology of Barnett and Lewis (1978), is an observation which does not belong to one of the target populations $G_1, \ldots, G_g$.

For a given $\mathbf{x}_j$, the Mahalanobis squared distance

$$D(\mathbf{x}_j, \bar{\mathbf{y}}_i; \mathbf{S}_i) \tag{2.6.1}$$

is computed for $i = 1, \ldots, g$. Then under the normal heteroscedastic model

(2.5.9),

$$c(m_i + 1, \nu_i + 1)D(\mathbf{x}_j, \bar{\mathbf{y}}_i; \mathbf{S}_i) \tag{2.6.2}$$

has a $F_{p,\nu_i+1}$ distribution where as before $\nu_i = m_i - p - 1$ and the function $c(.,.)$ is defined by (2.5.6). The situation here is more straightforward than with the assessment of typicality for a classified $\mathbf{y}_{ij}$, since $\mathbf{x}_j$ is an observation independent of $\bar{\mathbf{y}}_i$ and $\mathbf{S}_i$ $(i = 1, \ldots, g)$. An assessment of how typical $\mathbf{x}_j$ is of $G_i$ is given by $a_{i,j}$, the tail area to the right of the observed value of (2.6.2) under the $F_{p,\nu_i+1}$ distribution. For the homoscedastic model (2.5.1), $a_{i,j}$ is defined similarly, but where $\mathbf{S}_i$ is replaced by the pooled sample covariance matrix $\mathbf{S}$ in (2.6.1) and (2.6.2) and, as a consequence, $\nu_i$ is replaced by $\nu = m - g - p$ in (2.6.2). An observation $\mathbf{x}_j$ can be assessed as being atypical of the mixture if

$$a_j = \max_i a_{i,j} \leq \alpha, \tag{2.6.3}$$

where $\alpha$ is some specified threshold. The value of $\alpha$ depends on how the presence of apparently atypical observations is handled. If the aim is to undertake estimation with all atypical observations deleted from the sample, then $\alpha$ might be set at a conventional level, say $\alpha = 0.05$. On the other hand, the aim might be to eliminate only those observations assessed as being extremely atypical, so $\alpha$ might be set at 0.01 or 0.005. Protection against the possible presence of less extreme outliers can be provided by the use of robust $M$-estimators, to be discussed in Section 2.8. A third course of action might be to use redescending $M$-estimators to accommodate all discordant outliers and contaminants in the data, no matter how atypical. These matters are discussed in more detail in Section 2.8. Another approach for accommodating possible outliers within the estimation process might be to proceed as in Aitkin and Tunnicliffe Wilson (1980) and allow for another group (or groups) in the mixture model. The starting value(s) for the parameter vector(s) of the additional group(s) is based on a specification of one or more of the observations as outliers. An application of this approach is to be described in Section 3.3. More recently, Butler (1986) has examined the role of predictive likelihood inference in outlier theory.

Concerning the use of $a_j$ as a measure of typicality for the mixture of $\mathbf{x}_j$, it can be interpreted also as a $P$-value for a test of the compatibility of $\mathbf{x}_j$ with the mixture model (2.1.1) with normal component populations. For under the null hypothesis that $\mathbf{x}_j$ belongs to a normal mixture, the distribution of $a_j$ should be approximately uniform on the unit interval. If we consider the distribution of $a_j$ by conditioning first on $\mathbf{x}_j$ belonging to some $G_i$, say $G_t$, then $a_j$ is distributed conditionally as uniform on (0,1) if $a_{i,j}$ takes its maximum value at $i = t$. Hence the unconditional distribution of $a_j$ is precisely uniform on (0,1) if

$$\max_i a_{i,j}, \tag{2.6.4}$$

when viewed as defining an allocation rule for the assignment of $\mathbf{x}_j$ to one of $G_1, \ldots, G_g$, has a zero overall misallocation rate. Of course this is not so, but the overall misallocation rate of (2.6.4) will be small if the populations $G_i$ are well separated. For example, in the case of $g = 2$ normal populations with the same covariance matrix, Anderson (1958, 1984) showed that (2.6.4) defined the likelihood ratio rule, and Das Gupta (1965) subsequently showed it was minimax. Hence depending on the extent to which the $G_i$ are separated, we have approximately that

$$a_j \sim U(0, 1).$$

Note, that as $n \to \infty$ under the homoscedastic model, $a_j$ converges in probability to the area to the right of

$$\min_i D(\mathbf{x}_j, \mu_i; \Sigma) \tag{2.6.5}$$

under the $\chi_p^2$ distribution. Hawkins, Muller and ten Krooden (1982) noted that the null distribution of (2.6.5) is approximately $\chi_p^2$ for well separated populations. The statistic (2.6.5) arose as part of the attempt by Hawkins, Muller and ten Krooden (1982) to directly assess the fit of the mixture model in terms of the marginal distribution of $\mathbf{x}_j$ rather than in terms of its conditional distribution in each $G_i$ in the situation where the sample was completely unclassified. They observed that for a given $\boldsymbol{\mu}_i$,

$$D(\mathbf{x}_j, \boldsymbol{\mu}_i; \Sigma) \tag{2.6.6}$$

has a mixture of noncentral chi-squared distributions. One of their choices for the group mean to be used in (2.6.6) with a view to making its distribution close to a central $\chi_p^2$, was the mean of the group with the largest estimated mixing proportion. The statistic (2.6.5) was subsequently suggested as an approximate choice in the case of widely separated groups.

The monitoring of the unclassified observations $\mathbf{x}_1, \ldots, \mathbf{x}_n$ with respect to the classified data in terms of the tail area $a_{i,j}$ computed for each $i$ for a given $\mathbf{x}_j$, could have been expressed equivalently in terms of the area $1 - a_{i,j}$, which corresponds here to the value of the atypicality index for

$G_i$ of $\mathbf{x}_j$, as introduced by Aitchison and Dunsmore (1975). Their index of atypicality for $G_i$ of an observation $\mathbf{x}$ with arbitrary density function $f_i(\mathbf{x};\boldsymbol{\theta})$ in $G_i$ $(i = 1,\ldots,g)$ is defined by

$$\eta_i(\mathbf{x}) = \mathrm{pr}\{H_i(\mathbf{x}) \mid \mathbf{y} \in G_i\}$$

where

$$H_i(\mathbf{x}) = \{\mathbf{y} \mid f_i(\mathbf{y};\boldsymbol{\theta}) > f_i(\mathbf{x};\boldsymbol{\theta})\}$$

is the set of all observations more typical of $G_i$ than observation $\mathbf{x}$.

In order to provide an assessment of $\eta_i(\mathbf{x})$, we have to estimate the population densities $f_i(\mathbf{x};\boldsymbol{\theta})$. On the basis of the classified data $\mathbf{y}_{ij}$ $(i = 1,\ldots,g; j = 1,\ldots,m_i)$ under the normal heteroscedastic model (2.5.9), the predictive estimate of $f_i(\mathbf{x},\boldsymbol{\theta})$ obtained by a fully Bayesian approach using conventional improper prior distributions (Aitchison and Dunsmore, 1975, Chapter 2), is given by

$$\hat{\mathbf{f}}_i(\mathbf{x}) = f_{St}(\mathbf{x}; m_i - 1, \bar{\mathbf{y}}_i, (1 + 1/m_i)\mathbf{S}_i). \tag{2.6.7}$$

Here $f_{St}(\mathbf{x}; m_i, \bar{\mathbf{y}}_i, \mathbf{S}_i)$ denotes a $p$-dimensional Student-type density function

$$c(m_i)|m_i\mathbf{S}_i|^{-1/2}\left\{1 + D(\mathbf{x},\bar{\mathbf{y}}_i; m_i\mathbf{S}_i)\right\}^{(m_i+1)/2},$$

where

$$c(m_i) = \Gamma\left\{\tfrac{1}{2}(m_i + 1)\right\} \Big/ \left[\pi^{p/2}\Gamma\left\{\tfrac{1}{2}(m_i - p + 1)\right\}\right].$$

On forming and computing the probability of $H_i(\mathbf{x})$ as if the true density of $\mathbf{x}$ in $G_i$ were $\hat{\mathbf{f}}_i(\mathbf{x})$, the assessment of $\eta_i(\mathbf{x}_j)$ is given by

$$\hat{\eta}_i(\mathbf{x}_j) = 1 - a_{i,j}. \tag{2.6.8}$$

That is, $\hat{\eta}_i(\mathbf{x}_j)$, can be interpreted in a frequentist sense as one minus the level of significance associated with the test of the compatibility of $\mathbf{x}_j$ with $G_i$ on the basis of the Mahalanobis squared distance, or equivalently Hotelling's $T^2$; see Moran and Murphy (1979). The result (2.6.8) also holds under homoscedasticity for which the predictive estimate of $f_i(\mathbf{x};\boldsymbol{\theta})$ is given by

$$\hat{f}_i(\mathbf{x}) = f_{St}(\mathbf{x}; m - g, \bar{\mathbf{y}}_i, (1 + 1/m_i)\mathbf{S}). \tag{2.6.9}$$

So far as the estimation of $f_i(\mathbf{x};\boldsymbol{\theta})$ from a frequentist point of view, the so-called estimative method uses $f_i(\mathbf{x};\hat{\boldsymbol{\theta}})$, where $\hat{\boldsymbol{\theta}}$ is the likelihood estimate of $\boldsymbol{\theta}$ computed from the available data. In the role of providing estimates of the density functions for use in forming estimates of the posterior probabilities defined by (1.4.3), the predictive is preferable to the estimative method as it generally gives less extreme estimates of the posterior probabilities; see, for example, Aitchison, Habbema and Kay (1977), Desu and Geisser (1973), McLachlan (1979), and Rigby (1982). For normal heteroscedastic component densities, the estimative method takes $f_i(\mathbf{x};\boldsymbol{\theta})$ to be normal with mean $\bar{\mathbf{y}}_i$ and covariance matrix $\{(m_i-1)/m_i\}\,\mathbf{S}_i$, although $\mathbf{S}_i$ is usually used instead of the latter. On forming and computing the probability of $H_i(\mathbf{x})$ as if the true density in $G_i$ were equal to the estimative version, the assessment of $\eta_i(\mathbf{x}_j)$ is the area to the left of $D(\mathbf{x}_j,\bar{\mathbf{y}}_i;\mathbf{S}_i)$ under the $\chi_p^2$ distribution. This is equivalent to taking $\bar{\mathbf{y}}_i=\boldsymbol{\mu}_i$ and $\mathbf{S}_i=\boldsymbol{\Sigma}_i$ in considering the distribution of $D(\mathbf{x}_j,\bar{\mathbf{y}}_i;\mathbf{S}_i)$, and so gives a cruder assessment than the predictive approach.

## 2.7 ASSESSMENT OF NORMALITY AND TYPICALITY: UNCLASSIFIED DATA

In the situation where the sample is completely unclassified, as in the usual cluster analysis setting where there is no genuine group structure, it is a difficult task to assess the fit of a mixture model. As noted previously, we can assess the fit of a mixture model by proceeding conditionally and considering the distribution of $\mathbf{x}$ given the event that $\mathbf{x}$ comes from the $i$th component $G_i$ $(i=1,\ldots,g)$. For a mixture of normal distributions, this approach leads to the consideration of the null hypothesis

$$H_0:\mathbf{x}\sim N(\boldsymbol{\mu}_i,\boldsymbol{\Sigma})\qquad\text{in}\qquad G_i\ (i=1,\ldots,g). \tag{2.7.1}$$

It has been seen in Section 2.5 that Hawkins' (1980) test is a very useful way of assessing (2.7.1) where there are classified data available from each $G_i$. In the absence of classified data, Hawkins, Muller and ten Krooden (1982) suggested that the data $\mathbf{x}_1,\ldots,\mathbf{x}_n$ be clustered first according to an application of the mixture likelihood approach and then Hawkins' test applied to the resulting clusters as if they represented a correct grouping of the data relative to some externally existing populations.

Accordingly, we let $\tilde{\mathbf{x}}_{ij}$ $(j=1,\ldots,\tilde{m}_i)$ denote the $\tilde{m}_i$ observations put in the $i$th cluster $(i=1,\ldots,g)$ by the mixture likelihood approach applied under the assumption of a mixture of normal but heteroscedastic distributions. That is, the $\tilde{\mathbf{x}}_{ij}$ are those $\mathbf{x}_j$ for which $\hat{z}_{ij}=1$, where $\hat{z}_{ij}=1$ if the estimated posterior probabilities satisfy $\hat{\tau}_{ij}>\hat{\tau}_{tj}$ $(t=1,\ldots,g;t\neq i)$, and 0 otherwise. With this notation we can express $\tilde{m}_i$ and the sample mean and covariance matrix, $\bar{\tilde{\mathbf{x}}}_i$ and $\widetilde{\mathbf{S}}_i$, of the $\tilde{\mathbf{x}}_{ij}$ $(j = 1, \ldots, \tilde{m}_i)$ as

$$\tilde{m}_i = \sum_{j=1}^{n} \hat{z}_{ij},$$

$$\bar{\tilde{\mathbf{x}}}_i = \sum_{j=1}^{n} \hat{z}_{ij}\mathbf{x}_j / \tilde{m}_i,$$

and

$$\widetilde{\mathbf{S}}_i = \sum_{j=1}^{n} \hat{z}_{ij}(\mathbf{x}_j - \bar{\tilde{\mathbf{x}}}_i)(\mathbf{x}_j - \bar{\tilde{\mathbf{x}}}_i)'/(\tilde{m}_i - 1)$$

for $i = 1, \ldots, g$, and

$$\widetilde{\mathbf{S}} = \sum_{i=1}^{g} (\tilde{m}_i - 1)\widetilde{\mathbf{S}}_i/(n - g).$$

Hawkins' (1980) test is applied now to the clustered data $\tilde{\mathbf{x}}_{ij}$ and the results interpreted in the same manner as with its application to the classified observations $\mathbf{y}_{ij}$, as described in Section 2.5.

It is self-serving to cluster the data under the assumption of normality for the component distributions when one is subsequently to test for it, but it is brought about by the limitations on such matters as hypothesis testing in a cluster analysis framework. However, Hawkins, Muller and ten Krooden (1982) reported that, although this approach to the assessment of (2.7.1) in a cluster analysis context is rather crude, they have found it to work fairly well in practice; indeed, to the extent that they suggest there is little incentive to develop more sophisticated procedures.

We proceed now to consider the detection of atypical observations on the basis that the above assessment concludes that a mixture of normals adequately fits the data. A measure of typicality for the unclassified data is provided by the $\tilde{a}_{ij}$, the tail areas which are associated with the $\tilde{\mathbf{x}}_{ij}$ in the same way as the $a_{ij}$ were associated with the classified observations $\mathbf{y}_{ij}$ in the previous section. That is, $\tilde{a}_{ij}$ is the area to the right of (2.5.7) or (2.5.10) depending on whether homoscedasticity is adopted or not, but with $\bar{\mathbf{y}}_i$, $\mathbf{S}_i$, $\mathbf{S}$, $m_i$, and $\nu_i$ replaced where appropriate by $\bar{\tilde{\mathbf{x}}}_i$, $\widetilde{\mathbf{S}}_i$, $\widetilde{\mathbf{S}}$, $\tilde{m}_i$, and $\tilde{\nu}_i = \tilde{m}_i - p - 1$ respectively.

Under normal component distributions, the $a_{ij}$ were seen in Section 2.5 to be distributed uniformly on (0,1), but this will only be approximately true for the $\tilde{a}_{ij}$, as the $\tilde{\mathbf{x}}_{ij}$ $(j = 1, \ldots, \tilde{m}_i)$ constitute a cluster rather than an independent sample from $G_i$ with no misclassified observations $(i = 1, \ldots, g)$. An observation $\tilde{\mathbf{x}}_{ij}$ is regarded as atypical of $G_i$, and hence of the mixture due to the clustered nature of the $\tilde{\mathbf{x}}_{ij}$, if

$$\tilde{a}_{ij} \leq \alpha, \tag{2.7.2}$$

where $\alpha$ is the specified threshold.

We now relate (2.7.2) back to a given observation $\mathbf{x}_j$ in the unclustered sample. Suppose that after a clustering of the sample according to the mixture likelihood approach applied for a mixture of normal distributions, $\mathbf{x}_j$ is relabeled as, say $\tilde{\mathbf{x}}_{tk}$ for some $k (1 \leq k \leq \tilde{m}_t)$; that is, the estimated posterior probability of $\mathbf{x}_j$, $\hat{\tau}_{ij}$, is greatest with respect to $G_t$, and so $\hat{z}_{ij} = 1$ $(i = t)$ and $0 (i \neq t)$. Thus $\tilde{a}_{tk}$, denoted henceforth as $\tilde{a}_j$, can be written as

$$\tilde{a}_j = \sum_{i=1}^{g} \hat{z}_{ij} \tilde{a}_{i,j} \tag{2.7.3}$$

where $\tilde{a}_{i,j}$ corresponds to the typicality index $a_{i,j}$ for $\mathbf{x}_j$ assessed on the basis of the classified data $\mathbf{y}_{ij}$ (see previous section). Here now in the absence of classified data, $\tilde{a}_{i,j}$ denotes the area to the right of

$$\frac{(\tilde{\nu}_i \tilde{m}_i / p) D(\mathbf{x}_j, \bar{\tilde{\mathbf{x}}}_i; \widetilde{\mathbf{S}}_i)}{(\tilde{\nu}_i + p)(\tilde{m}_i - 1) - \tilde{m}_i D(\mathbf{x}_j, \bar{\tilde{\mathbf{x}}}_i; \widetilde{\mathbf{S}}_i)} \tag{2.7.4}$$

under the $F_{p,\tilde{\nu}_i}$ distribution; similarly, for homoscedastic components by substituting $\nu = m - g - p$ and $\widetilde{\mathbf{S}}$ for $\tilde{\nu}_i$ and $\widetilde{\mathbf{S}}_i$ respectively where they occur. Note that in general $\bar{\tilde{\mathbf{x}}}_i$ and $\widetilde{\mathbf{S}}_i$ provide inconsistent estimates of $\boldsymbol{\mu}_i$ and $\boldsymbol{\Sigma}_i$ respectively, as they are formed from the unclassified data $\mathbf{x}_j$ using the zero-one weights $\hat{z}_{ij}$ instead of the fractional weights given by the estimated posterior probabilities $\hat{\tau}_{ij}$. Therefore, it may be worthwhile to replace $\bar{\tilde{\mathbf{x}}}_i$ and $\widetilde{\mathbf{S}}_i$ by their likelihood estimates $\hat{\boldsymbol{\mu}}_i$ and $\widehat{\boldsymbol{\Sigma}}_i$ obtained by the mixture likelihood approach. If a normal heteroscedastic model is adopted for the component distributions, then the $\hat{\boldsymbol{\mu}}_i$ and $\widehat{\boldsymbol{\Sigma}}_i$ will already be available from the initial clustering of the $\mathbf{x}_j$ which has to be performed under this model before Hawkins' test is applied. However, if a homoscedastic model seems appropriate, then we would have to reapply the mixture likelihood approach under this additional assumption in order to obtain the $\hat{\boldsymbol{\mu}}_i$ and $\widehat{\boldsymbol{\Sigma}}$. In the subsequent work, it is assumed that the $\bar{\tilde{\mathbf{x}}}_i$ and $\widetilde{\mathbf{S}}_i$ are replaced by the $\hat{\boldsymbol{\mu}}_i$ and $\widehat{\boldsymbol{\Sigma}}_i$ in (2.7.4).

It is of interest to compare the typicality measure of $\mathbf{x}_j$, $\tilde{a}_j$, given by (2.7.3) with that based on the area to the right of the statistic

$$\min_i D(\mathbf{x}_j, \hat{\boldsymbol{\mu}}_i; \widehat{\Sigma}_i) \tag{2.7.5}$$

under the $\chi_p^2$ distribution. As discussed in the previous section, (2.7.5) was put forward by Hawkins, Muller and ten Krooden (1982) as an approximation for assessing the fit of a normal mixture with widely separated components. For sufficiently large $n$, $\tilde{a}_j$ can be approximated by the area to the right of

$$\sum_{i=1}^{g} \hat{z}_{ij} D(\mathbf{x}_j, \hat{\boldsymbol{\mu}}_i; \widehat{\Sigma}_i) \tag{2.7.6}$$

under the $\chi_p^2$ distribution.

By a similar argument as given in Section 2.6 on the distribution of the homoscedastic version of (2.7.5) under the null hypothesis of a normal mixture model, it follows that the null distribution of (2.7.6) is also approximately $\chi_p^2$ for well separated groups. It is assumed here that $n$ is sufficiently large for the variation in $\hat{\boldsymbol{\mu}}_i$ and $\widehat{\Sigma}_i$ to be ignored. Contrasting the statistic (2.7.5) with (2.7.6), it can be seen that the latter is based on the Mahalanobis squared distance between $\mathbf{x}_j$ and the sample mean of the group to which it has the highest estimated posterior probability of belonging; that is, the reference group for $\mathbf{x}_j$ is selected on the basis of the sample version of the optimal rule for the allocation of $\mathbf{x}_j$ to one of $G_1, \ldots, G_g$. Under homoscedasticity, the values of the statistics (2.7.5) and (2.7.6) are the same when they are formed with the mixing proportions estimated to be equal. For unequal estimates of the mixing proportions, and moreover under heteroscedasticity, examples can be constructed where for some $\mathbf{x}_j$, $D(\mathbf{x}_j, \hat{\boldsymbol{\mu}}_i; \widehat{\Sigma}_i)$ will actually be the largest in the group to which it has the highest estimated posterior probability of belonging. But in general (2.7.5) and (2.7.6) should coincide for most $\mathbf{x}_j$, the extent of their agreement depending on the degree of separation among the groups, the disparity of the estimates of the mixing proportions, and the severity of the heteroscedasticity.

## 2.8 ROBUST ESTIMATION FOR MIXTURE MODELS

We now consider for the mixture model a robust estimation procedure whereby observations assessed as atypical of a component or the mixture itself are automatically given reduced weight in the computation of the estimate of $\phi$. We shall present here the defining equations for the robust estimate of $\phi$ where, in addition to the unclassified data $\mathbf{x}_1, \ldots, \mathbf{x}_n$, there are $m_i$ observations $\mathbf{y}_{ij}$ ($j = 1, \ldots, m_i$) known to come from $G_i$ ($i = 1, \ldots, g$), and $m = \sum_{i=1}^{i=g} m_i$. However, the presence of the classified data is not crucial to the process and their presence can be ignored in the defining equations to obtain the appropriate equations for a completely unclassified sample.

Before we proceed to present the defining equations, we shall give a brief outline of their development by first looking at robust $M$-estimation (Huber, 1964) of a location parameter vector and scatter matrix in a non-mixture context by conditioning on a particular component of the mixture, say $G_i$. In the multivariate case, robust $M$-estimation was developed by taking an elliptically symmetric density and then associating it with a contaminated normal density; see Maronna (1976) and Huber (1977 and 1981, Chapter 8). To this end, suppose for the present that $f_i(\mathbf{x}; \boldsymbol{\theta})$ is a member of the family of $p$-dimensional elliptically symmetric densities,

$$|\boldsymbol{\Sigma}_i|^{-1/2} f_S \{D(\mathbf{x}, \boldsymbol{\mu}_i; \boldsymbol{\Sigma}_i)\}, \tag{2.8.1}$$

which can be generated by a nonsingular transformation of $\mathbf{x}$ from the family of spherically symmetric densities $f_S(\|\mathbf{x}\|)$; $\|\mathbf{x}\|$ denotes the Euclidean norm of $\mathbf{x}$. Under (2.8.1), $\boldsymbol{\Sigma}_i$ is a scalar multiple of the covariance matrix of $\mathbf{x}$. Further references on this approach may be found in Collins and Wiens (1985).

The $M$-estimates of $\boldsymbol{\mu}_i$ and $\boldsymbol{\Sigma}_i$ proposed by Maronna (1976) are defined by the equations

$$\sum_{j=1}^{n} u_1(\hat{d}_{ij})(\mathbf{y}_{ij} - \hat{\boldsymbol{\mu}}_i) = \mathbf{0} \tag{2.8.2}$$

and

$$\sum_{j=1}^{n} u_2(\hat{d}_{ij}^2)(\mathbf{y}_{ij} - \hat{\boldsymbol{\mu}}_i)(\mathbf{y}_{ij} - \hat{\boldsymbol{\mu}}_i)'/n = \tilde{\boldsymbol{\Sigma}}_i, \tag{2.8.3}$$

where for convenience $D(\mathbf{y}_{ij}, \boldsymbol{\mu}_i; \boldsymbol{\Sigma}_i)$ is written as $d_{ij}^2$ and where $u_1(s)$ and $u_2(s)$ are nonnegative weight functions. Under fairly general conditions on $u_1(s)$ and $u_2(s)$, Maronna (1976) established the existence and uniqueness of the solution of (2.8.2) and (2.8.3), and showed that it is consistent and asymptotically normal under (2.8.1). One of the conditions on the weight functions, that $su_1(s)$ and $su_2(s)$ are bounded, ensures that these estimates of $\boldsymbol{\mu}_i$ and $\boldsymbol{\Sigma}_i$ will be robust. It can be seen that (2.8.2) and (2.8.3) give the likelihood estimates of $\boldsymbol{\mu}_i$ and $\boldsymbol{\Sigma}_i$ if

$$u_1(s) = -s^{-1}\partial \log f_S(s)/\partial s$$

and $u_2(s^2) = u_1(s)$, for $s > 0$.

More generally to (2.8.3), Huber (1977 and 1981, page 213) observed that an affinely invariant estimate of $\boldsymbol{\Sigma}_i$ can be defined by

$$\sum_{j=1}^{n} u(\hat{d}_{ij})(\mathbf{y}_{ij} - \hat{\boldsymbol{\mu}}_i)(\mathbf{y}_{ij} - \hat{\boldsymbol{\mu}}_i)' \Big/ \sum_{j=1}^{n} v(\hat{d}_{ij}) \tag{2.8.4}$$

for arbitrary functions $u$ and $v$, and noted it was "particularly attractive" to take $u \equiv v$. The $M$-estimate of a covariance matrix in high dimensions has a low breakdown point $\alpha_0$ (which, roughly speaking, is the limiting proportion of bad outliers which can be tolerated by the estimate). For example, Maronna (1976) established that $\alpha_0 \leq 1/(p+1)$ for (2.8.3), where $u_2$ is monotone increasing and $u_2(0) = 0$, while Huber (1977) showed that this could be improved only to $\alpha_0 \leq 1/p$ in the more general framework (2.8.4).

In the above equations for the estimates of $\boldsymbol{\mu}_i$ and $\boldsymbol{\Sigma}_i$, the idea is to give full unit weight to those observations assessed as coming from the main body of the data, but to give reduced weight to any observation $\mathbf{y}_{ij}$ if its estimated Mahalanobis distance from $\boldsymbol{\mu}_i$, $\hat{d}_{ij}$, is sufficiently large for $\mathbf{y}_{ij}$ to be considered atypical of $G_i$. One proposal by Maronna (1976) for the weight function $u_1(s)$ in (2.8.2) is

$$u_1(s) = \psi(s)/s,$$

where $\psi(s)$ is Huber's (1964) $\psi$-function defined by

$$\psi(s) = \begin{cases} s, & |s| \leq k_1(p), \\ k_1(p), & |s| > k_1(p), \end{cases} \tag{2.8.5}$$

for an appropriate choice of the "tuning" constant $k_1(p)$, which is written here as a function of $p$ to emphasize its dependence on the dimension of $\mathbf{x}$. For this choice of $u_1(s)$, we have that (2.8.2) is the likelihood equation for $\boldsymbol{\mu}_i$ if the density of $\mathbf{x}$ is given by (2.8.1) where, up to a normalizing constant,

$$\log f_S(s) = \begin{cases} -\frac{1}{2}s^2, & |s| \le k_1(p), \\ \frac{1}{2}\{k_1(p)\}^2 - k_1(p)|s|, & |s| > k_1(p). \end{cases} \tag{2.8.6}$$

An associated choice for a weight function in the estimation of $\boldsymbol{\Sigma}_i$ is to take

$$u(s) = v(s) = \{u_1(s)\}^2$$

in (2.8.4) to give

$$\widehat{\boldsymbol{\Sigma}}_i = \sum_{j=1}^{n} \{u_1(\hat{d}_{ij})\}^2 (\mathbf{y}_{ij} - \hat{\boldsymbol{\mu}}_i)(\mathbf{y}_{ij} - \hat{\boldsymbol{\mu}}_i)' \Bigg/ \sum_{j=1}^{n} \left\{u_1(\hat{d}_{ij})\right\}^2 . \tag{2.8.7}$$

Note that if $u_1(\hat{d}_{ij})$ and not its square were used as a weight in (2.8.7), then the influence of grossly atypical observations would not be bounded.

The value of the tuning constant $k_1(p)$ in Huber's $\psi$-function (2.8.5) depends on the amount of contamination in the data. In practice, there is usually no prior knowledge about the extent of the contamination, and so $k_1(p)$ is chosen to give estimators with reasonable performances over a range of situations. In the univariate case, $k_1(1)$ is generally taken to be between 1 and 2. For $p > 1$, Devlin, Gnanadesikan and Kettenring (1981) took $k_1(p)$ to be the square root of the 90th percentile of the $\chi_p^2$ distribution, $\sqrt{\chi^2_{p,0.9}}$, since $\hat{d}^2_{ij}$ is asymptotically $\chi^2_p$ under normality. Campbell (1984, 1985) recommends $k_1(1) = 2$ and that in the multivariate case, $k_1(p)$ be computed corresponding to $k_1(1) = 2$ by taking $d^2_{ij}$ to be $\chi^2_p$ and then using the approximation of Wilson and Hilferty (1931),

$$\chi^2_{p,\alpha} \approx p\left\{1 - 2/9p + (2/9p)^{1/2} q_\alpha\right\}^3, \tag{2.8.8}$$

for the $\alpha$th quantile of the $\chi^2_p$ distribution. In (2.8.8), $q_\alpha$ denotes the $\alpha$th quantile of the $N(0,1)$ distribution. Since $k_1(1) = 2$ is approximately equal to the square root of $\chi^2_{1,0.95}$, this leads to

$$k_1(p) = \left[p\left\{1 - 2/9p + (2/9p)^{1/2} 1.645\right\}^3\right]^{1/2}, \tag{2.8.9}$$

on setting $k_1(p) = \sqrt{\chi^2_{p,0.95}}$ and then approximating the latter according to (2.8.8).

In the case of small $m_i$, the $\chi_p^2$ approximation to the distribution of $\hat{d}_{ij}^2$ may be fairly crude, so we might consider basing the tuning constant $k_1(p)$ on the $F_{p,\nu_i}$ distribution for

$$\{(\nu_i m_i/p)\hat{d}_{ij}^2\}/\{(\nu_i + p)(m_i - 1) - m_i \hat{d}_{ij}^2\} \tag{2.8.10}$$

where, as in the derivation of this result under normality in Section 2.5, $\nu_i = m_i - p - 1$. This result is only approximate here since $\hat{\boldsymbol{\mu}}_i$ and $\widehat{\boldsymbol{\Sigma}}_i$ do not have the same sampling distribution as the sample mean and covariance matrix, $\bar{\mathbf{y}}_i$ and $S_i$. Using (2.8.10), $k_1(p)$ is given by

$$k_1(p, m_i) = [\{p(\nu_i + p)(m_i - 1)F_{p,\nu_i;\alpha}\} / \{m_i(\nu_i + pF_{p,\nu_i;\alpha})\}]^{1/2}, \tag{2.8.11}$$

where $F_{p,\nu_i;\alpha}$ denotes the $\alpha$th quantile, say $\alpha = 0.95$, of the $F_{p,\nu_i}$ distribution.

As $m_i$ tends to infinity, $k_1(p, m_i)$ tends to $\sqrt{\chi_{p,\alpha}^2}$ and so, providing $m_i$ is not small relative to $p, k_1(p, m_i)$ for $\alpha = 0.95$ should be close to $k_1(p)$, as given by (2.8.9). For example, for $p = 4$, $k_1(4) = 3.075$, and $k_1(4, m_i) = 2.555$, 2.847, 2.930, and 2.969 at $m_i = 10$, 20, 30, and 40 respectively. Except for $m_i = 10$, which represents quite a small sample relative to $p, k_1(4, m_i)$ is reasonably close to $k_1(4)$.

If it is desired that observations extremely atypical of $G_i$ should have zero weight for values of $\hat{d}_{ij}$ above a certain level (rejection point), then a redescending $\psi$-function can be used, for example Hampel's (1973) piecewise linear function,

$$\psi(s) = \begin{cases} s, & |s| \le k_1(p), \\ k_1(p), & k_1(p) < |s| \le k_2(p), \\ k_1(p)\dfrac{\{k_3(p) - s\}}{\{k_3(p) - k_2(p)\}}, & k_2(p) < |s| \le k_3(p), \\ 0, & |s| > k_3(p), \end{cases} \tag{2.8.12}$$

where $k_1(p)$ is chosen in the same way as $k_1(p)$ in Huber's $\psi$-function (2.8.5). Care must be taken in choosing the remaining tuning constants $k_2(p)$ and $k_3(p)$ to ensure that $\psi$ does not descend too steeply, as cautioned by Huber (1981, Chapter 4) who also warns that redescending estimators are susceptible to underestimation of scale and that there may be multiple solutions. It can be seen that using Huber's nondescending $\psi$-function with manual rejection of grossly atypical observations beforehand is almost equivalent to the use of a redescending $\psi$-function, since either procedure removes very extreme observations. The difference lies in the treatment of the observations over which there is some doubt as to their retention in the data. With a redescending $\psi$-function, this step is carried out automatically by having $\psi$ redescend to zero from $k_2(p)$. For further discussion on the comparative properties of nondescending and redescending $\psi$-functions, the reader is referred to Goodall (1983), who also considers the various smooth versions subsequently proposed for Hampel's original $\psi$-function (2.8.12).

On the basis of extensive practical experience, Campbell (1984, 1985) advocates for convenience of use (2.8.12) with tuning constants $k_1(1) = 2$, $k_2(1) = 3$, and $k_3(1) = 5$ for $p = 1$. In the multivariate case he recommends defining $k_1(p)$ by (2.8.9) and $k_2(p)$ and $k_3(p)$ by replacing 1.645 with 2.8 and 5 respectively in the right-hand side of (2.8.9). With this choice $k_1(p)$ and $k_2(p)$ are approximately equal to the square root of $\chi^2_{p,0.95}$ and $\chi^2_{p,0.9974}$ respectively.

We now turn to our problem where we wish to construct robust estimates of the unknown parameters of the component densities of a mixture $G$ on the basis of the unclassified data $\mathbf{x}_1, \ldots, \mathbf{x}_n$ drawn from $G$. As remarked at the beginning of the section, we shall present the results for the case where there may be some classified data $\mathbf{y}_{ij}$ $(j = 1, \ldots, m_i)$ available from the component $G_i$ $(i = 1, \ldots, g)$. Analogously to writing the Mahalanobis distances for the classified data $\mathbf{y}_{ij}$ as $d^2_{ij}$, we shall write $D(\mathbf{x}_j, \boldsymbol{\mu}_i; \Sigma_i)$ as $d^2_{i,j}$ for the unclassified data $\mathbf{x}_j$.

The extension of robust $M$-estimation to the mixture model has been given by Campbell (1984, 1985) by taking a mixture of elliptical densities (2.8.1) and then associating them with contaminated normal component densities. The equations used here to define a robust estimate of $\phi$ are for $i = 1, \ldots, g$,

$$\hat{\boldsymbol{\mu}}_i = \frac{\left\{\sum_{j=1}^{m_i} u_1(\hat{d}_{ij})\mathbf{y}_{ij} + \sum_{j=1}^{n} \hat{\tau}_{ij} u_1(\hat{d}_{i,j})\mathbf{x}_j\right\}}{\left\{\sum_{j=1}^{m_i} u_1(\hat{d}_{ij}) + \sum_{j=1}^{n} \hat{\tau}_{ij} u_1(\hat{d}_{i,j})\right\}}, \tag{2.8.13}$$

$$\hat{\Sigma}_i = \left[\sum_{j=1}^{m_i} \left\{u_1(\hat{d}_{ij})\right\}^2 (\mathbf{y}_{ij} - \hat{\boldsymbol{\mu}}_i)(\mathbf{y}_{ij} - \hat{\boldsymbol{\mu}}_i)' + \sum_{j=1}^{n} \hat{\tau}_{ij} \left\{u_1(\hat{d}_{i,j})\right\}^2 (\mathbf{x}_j - \hat{\boldsymbol{\mu}}_i)(\mathbf{x}_j - \hat{\boldsymbol{\mu}}_i)'\right] \Big/ N_i, \tag{2.8.14}$$

and

$$\hat{\pi}_i = \left(m_i + \sum_{j=1}^{n} \hat{\tau}_{ij}\right) \Big/ (m+n) \tag{2.8.15}$$

or

$$\hat{\pi}_i = \sum_{j=1}^{n} \hat{\tau}_{ij}/n, \tag{2.8.16}$$

depending on whether the classified data $\mathbf{y}_{ij}$ were obtained by sampling from the mixture $G$ or not. In (2.8.14), it is proposed we take

$$N_i = \sum_{j=1}^{m_i} \left\{u_1(\hat{d}_{ij})\right\}^2 + \sum_{j=1}^{n} \hat{\tau}_{ij} \left\{u_1(\hat{d}_{i,j})\right\}^2$$

and that the normal density be used for each $f_i(\mathbf{x};\boldsymbol{\theta})$ in (1.4.3) in forming the posterior probabilities $\tau_{ij}$. Possible choices for the weight function, $u_1(s) = \psi(s)/s$, include those discussed above where the $\psi$-function may be defined by either (2.8.5) or (2.8.12). Note that (2.8.13) to (2.8.16) represent the likelihood equation for a mixture of elliptical densities (2.8.1) if

$$u_1(s) = -s^{-1}\partial \log f_S(s)/\partial s$$

and if in (2.8.14) we use $u_1$ and not its square as the weight function and take

$$N_i = \left(m_i + \sum_{j=1}^{n} \hat{\tau}_{ij}\right) \qquad (i = 1, \ldots, g).$$

Also, the posterior probabilities $\tau_{ij}$ must be formed from (1.4.3) using the elliptical densities (2.8.1).

We have proposed here in the equations (2.8.13) to (2.8.16) defining our robust estimates for the mixture distribution to take the component densities $f_i(\mathbf{x};\boldsymbol{\theta})$ in the formation of the posterior probabilities $\tau_{ij}$ to be normal rather than, say, of the form corresponding to the $\psi$-function used in defining the weight function $u_1$. Indeed, for redescending $\psi$-functions like (2.8.12) with finite rejection points there is no corresponding density function. For Huber's $\psi$-function there is a corresponding density function as given by (2.8.6). But it would appear that its use in forming the posterior probabilities $\tau_{ij}$ would lead to conservative estimates of the $\tau_{ij}$, and so to conservative estimates of the mixing proportions $\pi_i$. To demonstrate this we consider a mixture of $g = 2$ univariate distributions with common scale parameter $\sigma^2$. We focus on those values of $x$ which receive less than full weight with Huber's $\psi$-function (2.8.5) and which are in the tails of the mixture distribution, namely

$$(x - \hat{\mu}_1)/\hat{\sigma} > k_1 \qquad \text{or} \qquad (x - \hat{\mu}_2)/\hat{\sigma} < -k_1, \tag{2.8.17}$$

where $k_1(1)$ is written simply now as $k_1$; $\hat{\mu}_1$ is taken greater than $\hat{\mu}_2$ here. For normal component densities the estimated posterior probability $\hat{\tau}_{1j}$ can be expressed as

$$\begin{aligned}\operatorname{logit} \hat{\tau}_{1j} &= \log\{\hat{\tau}_{1j}/(1 - \hat{\tau}_{1j})\} \\ &= \left[\{x - \tfrac{1}{2}(\hat{\mu}_1 + \hat{\mu}_2)\}/\hat{\sigma}\right]\{(\hat{\mu}_1 - \hat{\mu}_2)/\hat{\sigma}\} \\ &\quad + \log(\hat{\pi}_1/\hat{\pi}_2)\end{aligned} \tag{2.8.18}$$

for all $x$. Now for elliptical component densities (2.8.1) with $f_S(s)$ given by (2.8.6), logit $\hat{\tau}_{1j}$ for $x$ satisfying (2.8.17) is given by (2.8.18), but where the term

$$\left[\{x - \tfrac{1}{2}(\hat{\mu}_1 + \hat{\mu}_2)\}/\hat{\sigma}\right] \tag{2.8.19}$$

is replaced by

$$k_1 \operatorname{sign}\left[\{x - \tfrac{1}{2}(\hat{\mu}_1 + \hat{\mu}_2)\}/\hat{\sigma}\right].$$

That is, the difference (2.8.19) is truncated at $\pm k_1$ in forming $\hat{\tau}_{1j}$, whereas the difference $(x - \hat{\mu}_i)/\hat{\sigma}$ is truncated at $\pm k_1$ in the computation of $\hat{\mu}_i$ and $\hat{\sigma}^2$. Since under normality the square of (2.8.19) is distributed asymptotically as $\chi_1^2(\frac{1}{4}\Delta^2)$, where $\frac{1}{4}\Delta^2 = \frac{1}{4}(\mu_1 - \mu_2)^2/\sigma^2$ is the noncentrality parameter, and $k_1^2$ is usually taken to be around the 95th percentile of the central $\chi_1^2$ distribution, it would appear that observations which are assessed as atypical of each component, and hence of the mixture, are censored too severely in the formation of the estimates of the posterior probabilities.

One consequence of using the normal density in the formation of the posterior probabilities of the $\tau_{ij}$ is that the effect on the mixing proportions of an observation atypical of the mixture is not diminished, although this effect is at most $1/n$ for any one observation. For example, although an observation $\mathbf{x}_j$ is atypical of each component, and hence of the mixture, it might have an estimated posterior probability $\hat{\tau}_{ij}$ close to 1 for some component $G_i$, say $G_t$, and so its contribution to $\hat{\pi}_t$ is close to the maximum input of $1/n$. Of course in practice it is anticipated that extremely atypical observations will have been removed from the data, either manually through the use of some rejection procedure applied in conjunction with Huber's $\psi$-function or automatically through the use of a redescending $\psi$-function like (2.8.12).

As with likelihood estimation for a mixture of normal distributions described in Section 2.1, the estimates of the $\boldsymbol{\mu}_i$, $\Sigma_i$, and the $\pi_i$ are computed iteratively, and the process is continued until successive estimates change by an arbitrarily small amount. The equations (2.8.13) to (2.8.16) have been presented for heteroscedastic components, but (2.8.14) and the subsequent remarks can be appropriately modified to reflect homoscedasticity.

# 3
# Applications of Mixture Models to Two-Way Data Sets

## 3.1 INTRODUCTION

In this chapter, we consider the application of finite mixture models to four real data sets each viewed in the form of a two-way array. The first two applications are concerned with the fitting of normal mixtures. In the first example on some hemophilia data, it will be seen that the adoption of a homoscedastic normal model in the presence of some heteroscedasticity can considerably influence the likelihood estimates, in particular of the mixing proportions, and hence the consequent clustering of the sample at hand. In the second example, the emphasis is on the role of the mixture model as a method for the identification of outliers in a single sample, as proposed by Aitkin and Tunnicliffe Wilson (1980).

The third and fourth examples to be presented are on the fitting of mixture models with discrete component distributions. The third example considers the clustering of cases of a rare disease, sudden infant death syndrome, on the basis of the number of cases observed for various counties in North Carolina. By fitting a mixture of two Poisson distributions to these occurrences, Symons, Grimson and Yuan (1983) clustered the counties into two groups corresponding to high and low risk counties. The fourth example is in the context of latent class analysis with the fitting of a mixture of multivariate Bernoulli variables which are taken to be independent within each component population. It concerns the work of Aitkin, Anderson and Hinde (1981), who adopted this model to cluster teachers into distinct styles on the basis of several binary variables describing teaching behavior.

## 3.2 CLUSTERING OF HEMOPHILIA DATA

We present here the analysis of a real data set which, although consisting of only two bivariate populations, illustrates the caution that needs to be exercised in practice with likelihood estimation for mixture models. The main part of this analysis is taken from Basford and McLachlan (1985d). The data set under consideration was taken from Habbema, Hermans and van den Broek (1974) where, in the context of genetic counseling, the question of discriminating between normal women and hemophilia A carriers was considered on the basis of two variables, $(x)_1 = \log_{10}$ (AHF activity) and $(x)_2 = \log_{10}$ (AHF-like antigen). Both variables are scaled up by 100 here to simplify the presentation of the results. We let $G_1$ be the population of noncarriers or normals and $G_2$ the population of carriers. Reference data containing $n_1 = 30$ observations on known noncarriers and $n_2 = 45$ observations on known obligatory carriers were available. These data points are plotted in Figure 1, where they are numbered 1 to $n_i$ within each population of origin $G_i$ $(i = 1, 2)$. In the subsequent work we shall refer to these observations as $\mathbf{x}_{ij}$ $(i = 1, 2; j = 1, \ldots, n_i)$ when considered as a reference set of known classification and as $\mathbf{x}_j$ $(j = 1, \ldots, n; n = n_1 + n_2 = 75)$ when treated as an unclassified sample. It is in this latter context that the data set is to be analysed here by fitting a mixture model under which the $n = 75$ observations $\mathbf{x}_j$ are taken to be the observed values of a random sample from a mixture of $G_1$ and $G_2$ in some unknown proportions $\pi_1$ and $\pi_2$. Note that the data set is to be analysed in a general context and not in its original setting where each observation $\mathbf{x}_j$ would have an associated prior probability of being a noncarrier, taken to be the genetic chance of being normal as ascertained from the pedigree of the individual.

The populations $G_1$ and $G_2$ are taken to be bivariate normal with means $\boldsymbol{\mu}_1$ and $\boldsymbol{\mu}_2$ and covariance matrices $\boldsymbol{\Sigma}_1$ and $\boldsymbol{\Sigma}_2$ respectively. The vector of unknown parameters $\phi$ is given by

$$\phi = (\boldsymbol{\pi}', \boldsymbol{\mu}_1', \boldsymbol{\mu}_2', \boldsymbol{\xi}_1', \boldsymbol{\xi}_2')',$$

where

$$\boldsymbol{\xi}_i = ((\boldsymbol{\Sigma}_i)_{11}, (\boldsymbol{\Sigma}_i)_{12}, (\boldsymbol{\Sigma}_i)_{22})' \qquad (i = 1, 2). \tag{3.2.1}$$

The results of Hawkins' (1981) test of

$$H_0 : \mathbf{x} \sim N(\boldsymbol{\mu}_i, \boldsymbol{\Sigma}) \qquad \text{in} \qquad G_i \quad (i = 1, 2)$$

for normality and homoscedasticity, are given in Table 2. They are for the application of Hawkins' test to the data in their known classified form.

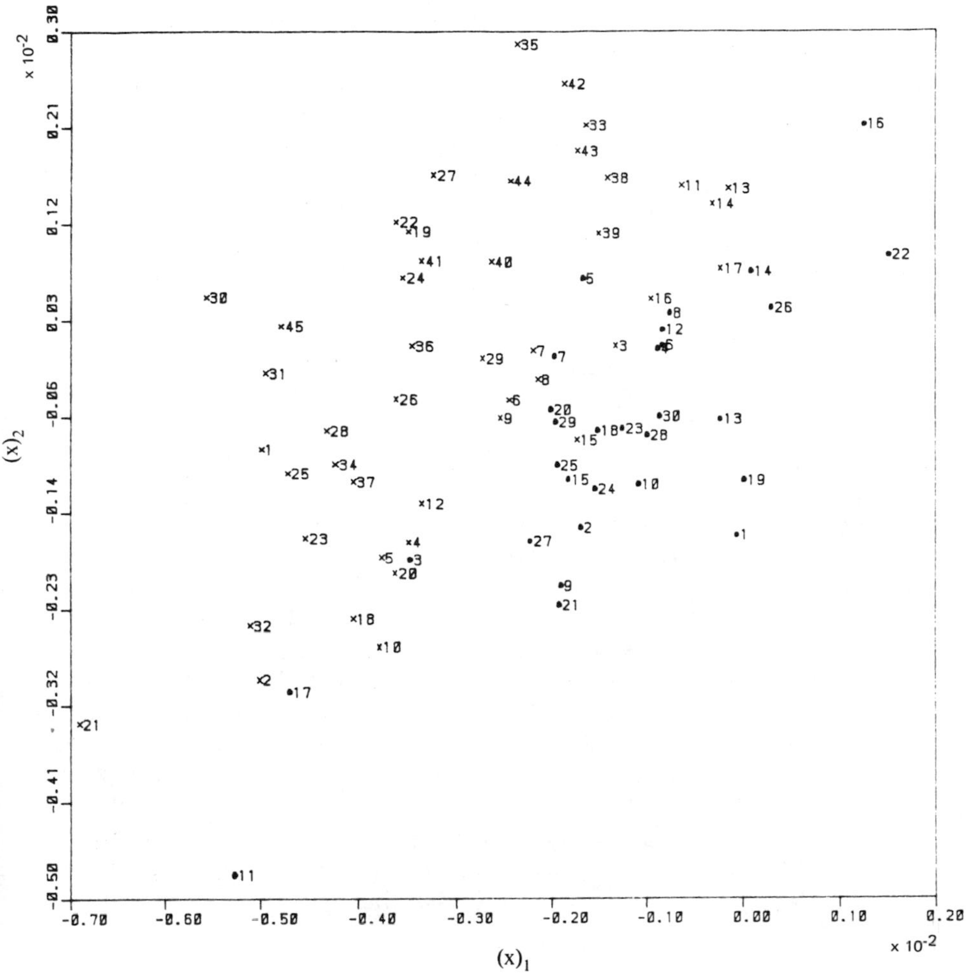


**FIGURE 1** Hemophilia data with • and × representing noncarriers and carriers respectively. (Data from Habbema, Hermans and van den Broek, 1974.)

That is, for each observation $\mathbf{x}_{ij}$ $(i = 1, 2; j = 1, \ldots, n_i)$, the tail area $a_{ij}$ to the right of the statistic corresponding to (2.5.7),

$$\frac{(\nu n_i/p) D(\mathbf{x}_{ij}, \bar{\mathbf{x}}_i; \mathbf{S})}{(\nu + p)(n_i - 1) - n_i D(\mathbf{x}_{ij}, \bar{\mathbf{x}}_i; \mathbf{S})} \tag{3.2.2}$$

was computed under the $F_{p,\nu}$ distribution, where $\bar{\mathbf{x}}_i$ and $\mathbf{S}_i$ denote the sample mean and covariance matrix of the $\mathbf{x}_{ij}$ $(j = 1, \ldots, n_i)$, $\mathbf{S}$ denotes the pooled version of the $\mathbf{S}_i$, and $\nu = n - g - p$. For this application, $p = 4$, $n = 75$, and $g = 2$. As explained in the presentation of Hawkins' (1981) test in Section 2.5, the Anderson-Darling statistic and its first two asymptotic $N(0,1)$ components for the $a_{ij}$ from $G_i$ and for the totality of the $a_{ij}$ are useful in interpreting qualitatively departures from $H_0$. From Table 2, the difference in sign and significance of the first components of the Anderson-Darling statistics for the individual populations indicate heteroscedasticity while the nonsignificance of the Anderson-Darling statistic and its components for the totality of the $a_{ij}$ gives a fair indication that $G_1$ and $G_2$ are bivariate normal.

An inspection of

$$\hat{\xi}_1 = (0.021, 0.015, 0.018)' \qquad \text{and} \qquad \hat{\xi}_2 = (0.024, 0.015, 0.024)',$$

where $\hat{\xi}_i$ is defined by replacing $\Sigma_i$ with $\mathbf{S}_i$ as given by (2.5.3) in the right-hand side of (3.2.1), shows that the only possible source of the indicated heteroscedasticity is the difference in the sample variance of the second component of $\mathbf{x}$, $(x)_2$, in $G_1$ and in $G_2$. The standard $F$ test for equality of variances applied to just the sample variances of $(x)_2$ gives a $P$-value of 0.20. However, this $P$-value drops considerably to 0.04 if the observation which has the smallest value of $(x)_1$ and of $(x)_2$ amongst those from $G_1$ (see Figure 1), is considered too discordant to be retained; the corresponding value of $a_{1j}$ $(j = 11)$ is 0.01. The values of $a_{ij}$ calculated using (3.2.2) are displayed in Table 3 for $i = 1, 2$ and $j = 1, \ldots, n_i$. An account of the sensitivity of the estimated covariance matrices to outliers may be found in Campbell (1980). It will be seen that the adoption of a heteroscedastic model at least with respect to $(x)_2$ is of some consequence here as the

**TABLE 2** Results of Hawkins' test for normality and homoscedasticity (applied to data in known classified form)

| Source | Anderson-Darling statistic | Components of Anderson-Darling statistic | |
|---|---|---|---|
| | | First | Second |
| $G_1$ | 3.14 | −2.24 | −1.70 |
| $G_2$ | 2.32 | 1.99 | 0.97 |
| Totality | 0.22 | 0.13 | −0.32 |

Source: Basford and McLachlan (1985d).

clustering of the sample corresponding to the likelihood estimate under this model is much more effective than that obtained under the assumption of homoscedasticity.

As seen in Section 2.5, Hawkins' (1981) test for multivariate normality and homoscedasticity is also useful for outlier detection. Other observations with values of $a_{ij}$ not greater than 0.05, in addition to the one noted above, are the first observation from $G_1(a_{1j} = 0.04, j = 1)$, the observation with the second smallest value of $(x)_1$ and of $(x)_2$ amongst those from $G_1(a_{1j} = 0.05, j = 17)$, and the two observations from $G_2$ with the smallest two values of $(x)_1$ in the entire sample ($a_{2j} = 0.03$, $j = 21$ and $a_{2j} = 0.04$, $j = 30$). The values of the $a_{ij}$ stated here, and reported in Table 3, were obtained after appropriately modifying (3.2.2) and the degrees of freedom in its $F$

**TABLE 3** Values of tail areas $a_{ij}$ under homoscedastic and heteroscedastic models

| | Homoscedastic | | Heteroscedastic | | | Homoscedastic | | Heteroscedastic | |
|---|---|---|---|---|---|---|---|---|---|
| $j$ | $a_{1j}$ | $a_{2j}$ | $a_{1j}$ | $a_{2j}$ | $j$ | $a_{1j}$ | $a_{2j}$ | $a_{1j}$ | $a_{2j}$ |
| 1 | 0.16 | 0.40 | 0.04 | 0.44 | 24 | 0.95 | 0.55 | 0.92 | 0.64 |
| 2 | 0.84 | 0.13 | 0.78 | 0.16 | 25 | 0.91 | 0.55 | 0.89 | 0.57 |
| 3 | 0.34 | 0.31 | 0.28 | 0.37 | 26 | 0.55 | 0.94 | 0.53 | 0.94 |
| 4 | 0.84 | 0.40 | 0.78 | 0.48 | 27 | 0.80 | 0.21 | 0.79 | 0.30 |
| 5 | 0.25 | 0.40 | 0.08 | 0.47 | 28 | 0.96 | 0.70 | 0.94 | 0.72 |
| 6 | 0.83 | 0.70 | 0.78 | 0.75 | 29 | 0.78 | 0.96 | 0.67 | 0.96 |
| 7 | 0.47 | 0.76 | 0.27 | 0.79 | 30 | 0.95 | 0.02 | 0.93 | 0.04 |
| 8 | 0.71 | 0.62 | 0.61 | 0.69 | 31 | | 0.24 | | 0.31 |
| 9 | 0.61 | 0.66 | 0.48 | 0.73 | 32 | | 0.24 | | 0.27 |
| 10 | 0.84 | 0.11 | 0.75 | 0.16 | 33 | | 0.32 | | 0.37 |
| 11 | 0.01 | 0.26 | 0.00 | 0.28 | 34 | | 0.74 | | 0.75 |
| 12 | 0.76 | 0.56 | 0.68 | 0.63 | 35 | | 0.05 | | 0.09 |
| 13 | 0.65 | 0.13 | 0.55 | 0.15 | 36 | | 0.90 | | 0.92 |
| 14 | 0.54 | 0.16 | 0.51 | 0.19 | 37 | | 0.74 | | 0.76 |
| 15 | 0.95 | 0.20 | 0.95 | 0.28 | 38 | | 0.47 | | 0.49 |
| 16 | 0.12 | 0.26 | 0.08 | 0.31 | 39 | | 0.57 | | 0.59 |
| 17 | 0.07 | 0.09 | 0.05 | 0.12 | 40 | | 0.80 | | 0.83 |
| 18 | 0.97 | 0.22 | 0.96 | 0.28 | 41 | | 0.56 | | 0.64 |
| 19 | 0.30 | 0.36 | 0.14 | 0.45 | 42 | | 0.16 | | 0.23 |
| 20 | 0.71 | 0.30 | 0.57 | 0.38 | 43 | | 0.41 | | 0.45 |
| 21 | 0.52 | 0.03 | 0.37 | 0.03 | 44 | | 0.46 | | 0.52 |
| 22 | 0.15 | 0.27 | 0.11 | 0.36 | 45 | | 0.18 | | 0.26 |
| 23 | 1.00 | 0.52 | 1.00 | 0.55 | | | | | |

distribution to allow for heteroscedasticity; that is, replacing the pooled sample covariance matrix $\mathbf{S}$ with $\mathbf{S}_i$ and $\nu$ by $\nu_i = n_i - p - 1$ $(i = 1, 2)$.

The above assessment of normality and homoscedasticity was carried out using the known population of origin of each observation. In the context where the sample is treated as being completely unclassified we have seen in Section 2.7 how an assessment can be made using a proposal of Hawkins, Muller and ten Krooden (1982). The sample is first clustered by fitting a mixture of normal distributions with unequal covariance matrices, and then Hawkins' (1981) test is applied to the resulting clusters as if they represent a correct partition of the data with respect to $G_1$ and $G_2$. The results of this method of assessment are displayed in Table 4. Like the results in Table 2 for Hawkins' test based on the known classification of the sample, they suggest that it is reasonable to take the component distributions to be bivariate normal, but not with equal covariance matrices, as some heteroscedasticity is indicated by the difference in sign and significance of the first components of the Anderson-Darling statistic for the individual populations.

Listed in Table 5 are the estimates of $\phi$ obtained as roots of the mixture likelihood equation under a bivariate normal, heteroscedastic model for the component distributions; $\hat{\phi}_C$ denotes the usual estimate of $\phi$ formed with knowledge of the true origin of the data, that is, using the log likelihood $L_C(\phi)$ for the complete data $\mathbf{X}$ and $\mathbf{Z}$. Application of the EM algorithm using a wide choice of starting values located two local maxima. It can be confirmed from Table 5 that the root corresponding to the larger of the two maxima, $\hat{\phi}_1$, is closer in Euclidean distance to $\hat{\phi}_C$. The boundary on which the posterior probability of $\mathbf{x}$ belonging to each $G_i$ is the same, $\tau_i(\mathbf{x}; \phi) = 0.5$, is plotted in Figure 2 for the various estimates of $\phi$. It can be observed that the use of $\hat{\phi}_1$ gives a better overall allocation of the data than $\hat{\phi}_2$, with three and twelve members from $G_1$ and $G_2$ respectively

**TABLE 4** Results of Hawkins' test for normality and homoscedasticity (applied to data in clustered form)

| Source | Anderson-Darling statistic | Components of Anderson-Darling statistic | |
|---|---|---|---|
| | | First | Second |
| $G_1$ | 1.99 | −1.87 | 0.17 |
| $G_2$ | 1.98 | 1.85 | −0.19 |
| Totality | 0.33 | −0.07 | −0.01 |

being misallocated. Everitt (1984a) has also provided examples (concerning mixtures of univariate normal densities with unequal variances), where similar likelihood values gave very different parameter estimates, but where in each instance the preferred solution corresponded to the largest of the local maxima located.

Application of the EM algorithm now under a normal but homoscedastic model for the component distributions,

$$\mathbf{x} \sim N(\mu_i, \Sigma) \quad \text{in} \quad G_i \ (i = 1, 2), \tag{3.2.3}$$

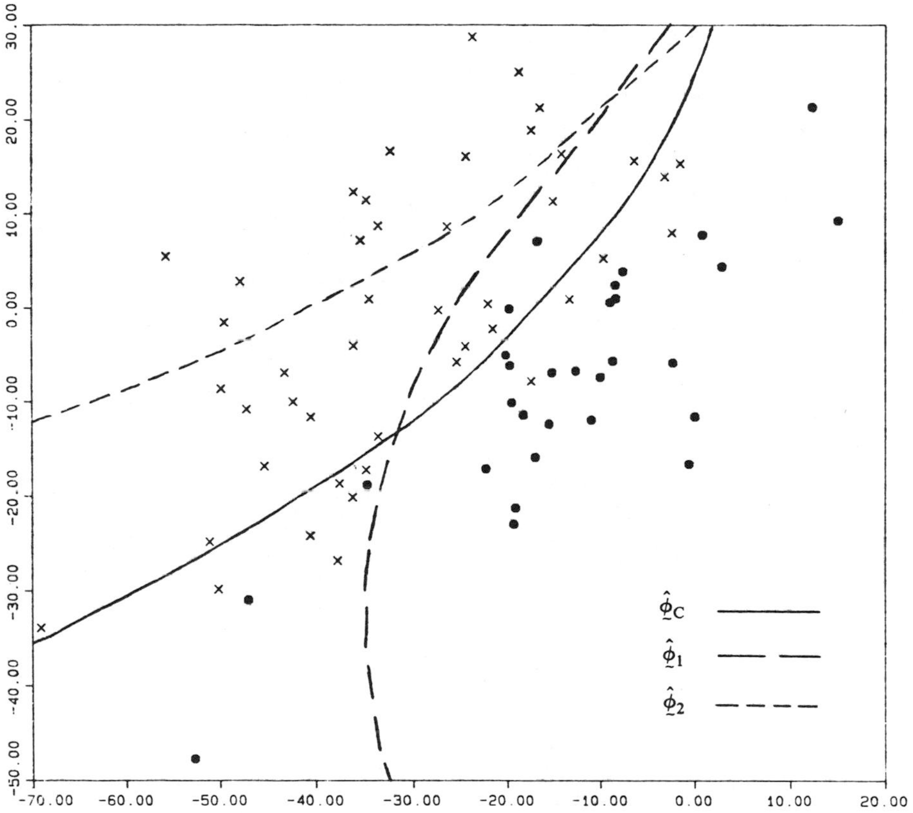


**FIGURE 2** Hemophilia data along with allocation boundaries, $\tau_i(\mathbf{x}; \phi) = 0.5$, for various estimates of $\phi$ under heteroscedasticity. (From Basford and McLachlan, 1985d.)

**TABLE 5** Estimates of $\phi$ under heteroscedasticity

| $\phi =$ | $(\pi_1,$ | $\pi_2,$ | $(\boldsymbol{\mu}_1)_1,$ | $(\boldsymbol{\mu}_1)_2,$ | $(\boldsymbol{\mu}_2)_1,$ | $(\boldsymbol{\mu}_2)_2,$ | $(\Sigma_1)_{11},$ | $(\Sigma_1)_{12},$ | $(\Sigma_1)_{22},$ | $(\Sigma_2)_{11},$ | $(\Sigma_2)_{12},$ | $(\Sigma_2)_{22})'$ | $L(\phi)$ |
|---|---|---|---|---|---|---|---|---|---|---|---|---|---|
| $\hat{\phi}_1 =$ | (0.503, | 0.497, | −11.4 | −2.4, | −36.4, | −4.5, | 111, | 65, | 123, | 160, | 150, | 321)′ | 77.04 |
| $\hat{\phi}_2 =$ | (0.814, | 0.186, | −21.9, | −7.1, | −32.4, | 12.4, | 305, | 165, | 184, | 148, | 87, | 81)′ | 76.80 |
| $\hat{\phi}_C =$ | (0.400, | 0.600, | −13.5, | −7.8, | −30.8, | −0.6 | 209, | 155, | 179, | 238, | 154, | 240)′ | — |

Source: Basford and McLachlan (1985d).

illustrates that the effectiveness of the clustering of the data corresponding to the maximum likelihood estimate can be well below that obtained with the likelihood estimate under the more appropriate heteroscedastic model. Let $\phi^{\dagger}$ denote the reduced form of $\phi$ with $\xi_1 = \xi_2$ and $\hat{\phi}_C^{\dagger}$ the estimate of $\phi^{\dagger}$ based on the true origin of the data. Three local maxima were located, and the corresponding estimates of $\phi^{\dagger}$ are displayed in Table 6. The allocation boundary, $\tau_i(\mathbf{x}; \hat{\phi}^{\dagger}) = 0.5$, is plotted in Figure 3 for the various estimates of $\phi^{\dagger}$. As $\hat{\phi}_1^{\dagger}$ corresponds to the largest of the local maxima located, it is apparently the maximum likelihood estimate which exists under (3.2.3). It can be seen from Table 6 that $\hat{\phi}_1^{\dagger}$ gives a poor assessment of the proportions in which $G_1$ and $G_2$ are represented in the sample and, as a consequence, it gives a worse allocation of the sample than $\hat{\phi}_2^{\dagger}$, with a total of 25 members, all from $G_2$, misallocated.

In a discriminant analysis context, where the estimates of the parameters are formed with the knowledge of the true origin of the data, it is well known that Fisher's linear discriminant function is fairly robust to departures from homoscedasticiy. Indeed, for the present example, it can be observed from Figures 2 and 3 that the use of $\hat{\phi}_C^{\dagger}$, equivalent to applying a linear discriminant function, misallocates only eleven members overall (four from $G_1$ and seven from $G_2$) compared to a total of fourteen misallocations (two from $G_1$ but twelve from $G_2$) with the use of $\hat{\phi}_C$, corresponding to a quadratic discriminant function. This robustness of the linear discriminant function suggests for the present data set, where any heteroscedasticity is limited to the second component of the observation vector, that there may be a root of the likelihood equation under (3.2.3) which yields a partition of the data comparable to that given by $\hat{\phi}_1$ under the heteroscedastic model. From Figure 4, it can be seen that the use of $\hat{\phi}_2^{\dagger}$, the root of the likelihood equation under (3.2.3) corresponding to the second largest of the maxima located, gives almost the same clustering of the data as $\hat{\phi}_1$. The only difference in the partitions is that one additional observation from $G_2$ is misallocated using $\hat{\phi}_2^{\dagger}$.

It is not difficult to understand why the local maximum of the mixture likelihood, corresponding to the root nearest to the estimate based on the true origin of the data, is not the largest when the condition of homoscedasticity is imposed. Under either model, this root of the likelihood equation gives a probabilistic clustering in which the variation of $(x)_2$ is substantially smaller in the cluster associated with $G_1$ than in the other associated with $G_2$, which is not consistent with homoscedasticity. In particular, this root provides estimates of $\pi_1$ and $\pi_2$ much closer to the actual proportions in which $G_1$ and $G_2$ occur in the sample than the maximum likelihood estimate under the homoscedastic model. Hence the apparent reason why this root leads to a better allocation of the sample than the maximum likeli-

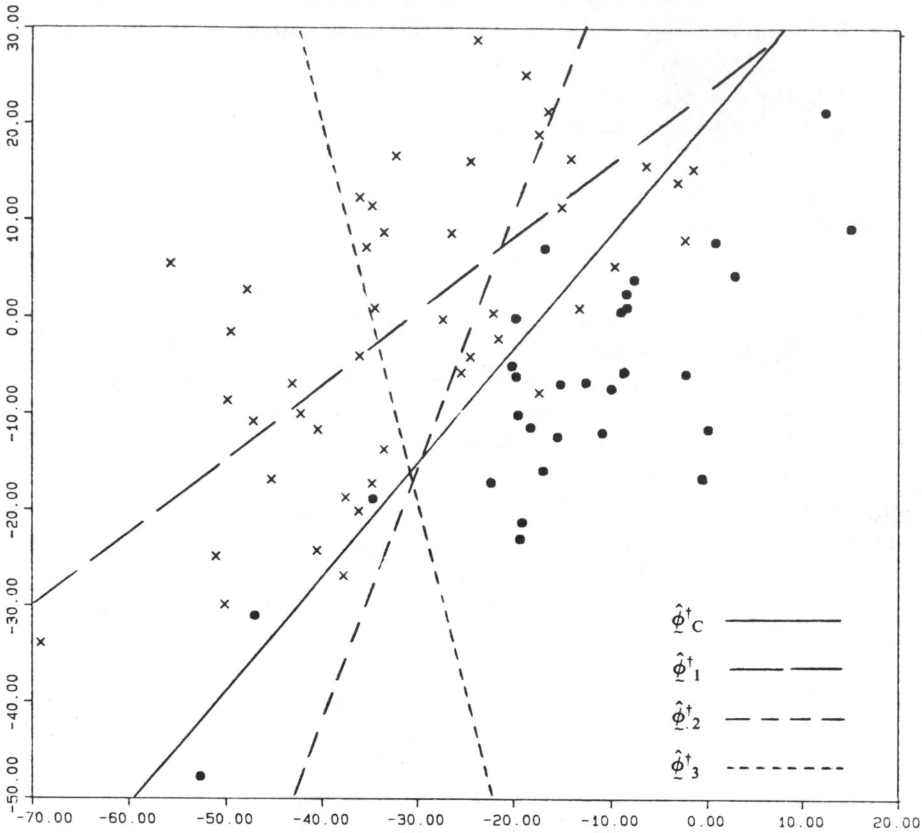


**FIGURE 3** Hemophilia data along with allocation boundaries, $\tau_i(\mathbf{x}; \phi^{\dagger}) = 0.5$, for various estimates of $\phi^{\dagger}$ under homoscedasticity. (From Basford and McLachlan, 1985d.)

hood estimate under the condition of homoscedasticity. Not surprisingly, $\hat{\phi}^{\dagger}_2$ may be obtained by starting the EM algorithm from $\hat{\phi}^{\dagger}_C$. The sensitivity of the iterative process to starting values, in particular for the mixing proportion parameter, is demonstrated by the fact that $\hat{\phi}^{\dagger}_1$ can be obtained by starting the EM algorithm from $\hat{\phi}^{\dagger}_C$ but with the value for $\pi_1$ increased

**TABLE 6** Estimates of $\boldsymbol{\phi}^{\dagger}$ under homoscedasticity

| $\boldsymbol{\phi}^{\dagger} =$ | $(\pi_1,$ | $\pi_2,$ | $(\boldsymbol{\mu}_1)_1,$ | $(\boldsymbol{\mu}_1)_2,$ | $(\boldsymbol{\mu}_2)_1,$ | $(\boldsymbol{\mu}_2)_2,$ | $(\Sigma)_{11},$ | $(\Sigma)_{12},$ | $(\Sigma)_{22})'$ | $L(\boldsymbol{\phi}^{\dagger})$ |
|---|---|---|---|---|---|---|---|---|---|---|
| $\hat{\boldsymbol{\phi}}_1^{\dagger} =$ | (0.716, | 0.284, | −20.6, | −8.0, | −32.1, | 7.9, | 265, | 158, | 171)′ | 75.00 |
| $\hat{\boldsymbol{\phi}}_2^{\dagger} =$ | (0.528, | 0.472, | −12.1, | −1.9, | −37.0, | −5.2, | 137, | 100, | 220)′ | 73.49 |
| $\hat{\boldsymbol{\phi}}_3^{\dagger} =$ | (0.681, | 0.319 | −15.3, | 1.2, | −42.0 | −13.5, | 138, | 35, | 175)′ | 73.29 |
| $\hat{\boldsymbol{\phi}}_C^{\dagger} =$ | (0.400, | 0.600, | −13.5, | −7.8, | −30.8, | −0.6, | 226, | 154, | 216)′ | — |

Source: Basford and McLachlan (1985d).

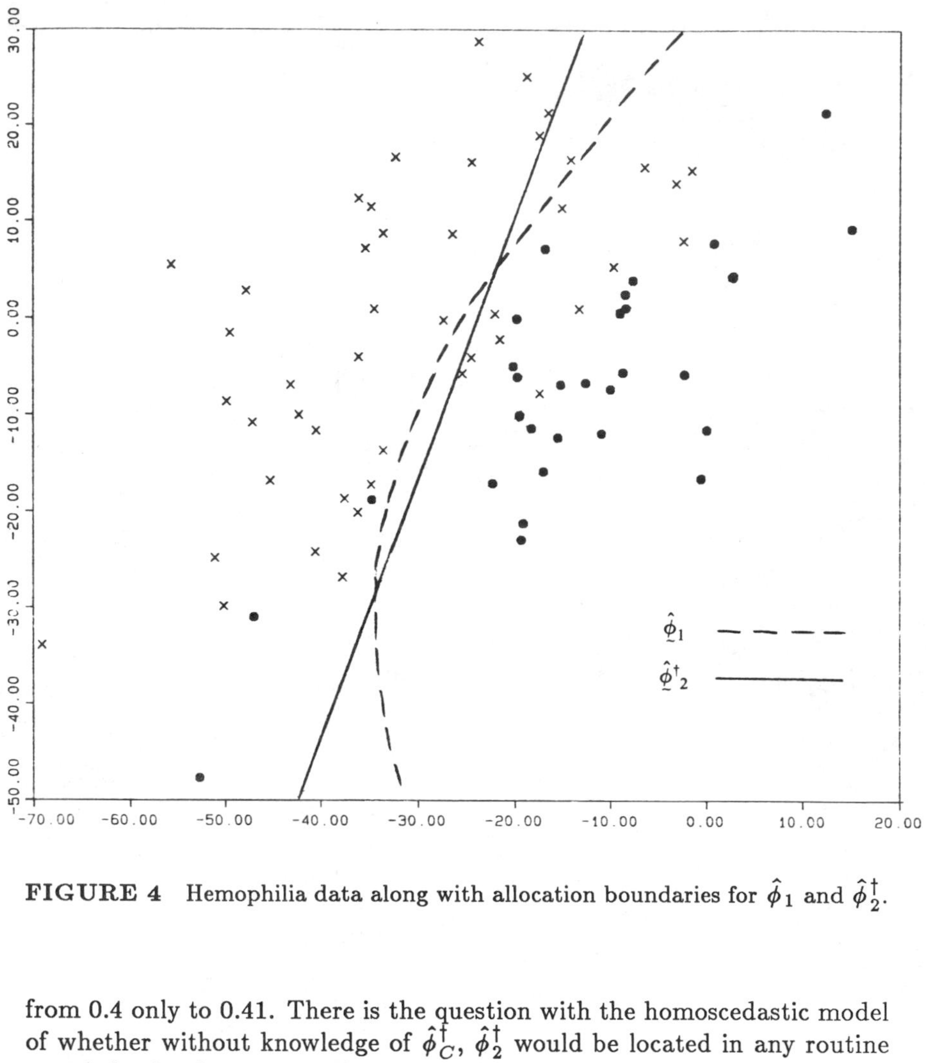


**FIGURE 4** Hemophilia data along with allocation boundaries for $\hat{\phi}_1$ and $\hat{\phi}^{\dagger}_2$.

from 0.4 only to 0.41. There is the question with the homoscedastic model of whether without knowledge of $\hat{\phi}^{\dagger}_C$, $\hat{\phi}^{\dagger}_2$ would be located in any routine search for local maxima. If attention is focused on only the first component of the data (there is little separation between $G_1$ and $G_2$ in the second component), it can be seen from Figure 3 that there is a distinct gap in the observations around $(x)_1 = -30$. A starting value for $\phi$ can be obtained then by partitioning the data in two groups according as to whether $(x)_1$ is greater or less than $-30$, and estimating $\phi$ on the basis that these two groups are correctly classified. The EM algorithm applied from this starting value leads to $\hat{\phi}^{\dagger}_2$.

As noted earlier after the application of Hawkins' (1981) test to this data set, one observation in particular, the one with the smallest value of $(x)_1$ and $(x)_2$ amongst those from $G_1$, that is, observation 11 from $G_1$, was identified as being highly atypical. However, it was found that its removal from the sample did not alter the result that under the homoscedastic model the maximum likelihood estimate did not give as good an allocation of the sample as the use of another root of the likelihood equation.

Although this data set is fairly simple in that it consists of only two bivariate populations, the latter are fairly close together. Indeed, the sample Mahalanobis squared distance between the two populations, $D(\bar{\mathbf{y}}_1, \bar{\mathbf{y}}_2; \mathbf{S})$, is only 4.57. It is of interest to see for this data set if the likelihood ratio test of

$$H_0 : g = 1 \text{ versus } H_1 : g = 2$$

would lead to the rejection of the null hypothesis of a single population. For the mixture model fitted above under the assumption of $g = 2$ bivariate normal distributions with unequal covariance matrices, $L(\hat{\phi}_1) = 77.04$, which is an increase of 5.3 over the maximum of the likelihood for a single bivariate normal population. As discussed in Section 1.10, the likelihood ratio criterion does not have its usual asymptotic distribution under $H_0$. With the approximation suggested by Wolfe (1971),

$$-2 \log \lambda \sim \chi_d^2 \tag{3.2.4}$$

under $H_0$, where the degrees of freedom $d$ is taken to be twice the difference in the number of parameters under $H_0$ and $H_1$, not including the mixing proportions. Here $d = 10$, and so with respect to the $\chi_{10}^2$ distribution, the value of 10.6 for $-2 \log \lambda$ is clearly not significant; its $P$-value is 0.39. As cautioned in Section 1.10, a result based on the approximation (3.2.4) should be used only as a guide to the choice of $g$.

To consider further the choice of $H_0$ versus $H_1$ on the basis of the log likelihood ratio, the bootstrap method is used now as described in Section 1.10 to provide a test of asymptotic size $\alpha$, where $\alpha$ is the specified level. For this test, $H_0$ is rejected at a nominal $\alpha$ level if $-2 \log \lambda$ evaluated for the original sample exceeds the $(1 - \alpha)(K + 1)$th smallest of $K$ bootstrap values subsequently replicated for this statistic, where $\alpha(K + 1)$ is an integer. Given the amount of computation involved, $K$ was limited to 19. Proceeding under $H_0$, 19 bootstrap samples of size $n = 75$ were generated from a single bivariate normal population. In this case the mean and covariance matrix of the latter are arbitrary since the null distribution of $\lambda$ does not depend on them. Hence the $K$ replications of $-2 \log \lambda$ are actually from the true distribution of this statistic, and so the size of the test is exactly

$\alpha$. The value of $\alpha$ in mind here is 0.05 (the smallest level possible with $K = 19$), but it turns out that $H_0$ would not be rejected even at the 0.3 level.

For each of the 19 bootstrap samples a mixture of $g = 2$ bivariate normal distributions with unequal covariance matrices was fitted, and the increase in the log likelihood over that for a single population calculated. As the value of 10.6 for $-2\log\lambda$ from the original sample was less than six of its 19 replications, $H_0$ would clearly not be rejected in favor of $H_1$. This conclusion of course is based on only $K = 19$ replications of $-2\log\lambda$. But even if an additional 80 replications had been obtained in the first instance so as to have provided a more powerful test with $K = 99$, the null hypothesis of a single population would still have been retained at the 0.05 level. This is because $-2\log\lambda$ for the original sample would not have exceeded its fifth largest replication from resampling.

## 3.3 OUTLIERS IN DARWIN'S DATA

To demonstrate the use of mixture models as a tool for the identification of outliers in a single sample or regression problems, Aitkin and Tunnicliffe Wilson (1980) analysed various data sets. We consider their analysis of Darwin's data on the differences in heights of 15 pairs of self-fertilized plants grown on the same plot, previously considered by Box and Tiao (1968). The ordered observations denoted here by $x_1, \ldots, x_{15}$ are listed in Table 7. They noted, from a simple probability plot, that the smallest two observations, $x_1$ and $x_2$, are clearly separate. Accordingly, they fitted a two-component mixture model. The component distributions were taken to be normal with the same variances. It was found it was not necessary to take the variances to be different.

The initial value of $\phi$ was obtained by setting the estimated posterior probability of belonging to the first component, $\hat{\tau}_{1j}$, equal to 1 ($j = 1, 2$) and equal to 0 ($j = 3, \ldots, 15$) in the univariate homoscedastic versions of (2.1.2) to (2.1.4). The final estimates of the posterior probabilities $\tau_{1j}$ confirmed the isolation of $x_1$ and $x_2$ with $\hat{\tau}_{1j} = 1$ ($j = 1, 2$) and $\hat{\tau}_{1j} \leq 0.001$ ($j = 3, \ldots, 15$). Reproduced here in Figure 5 from Aitkin and Rubin (1985)

**TABLE 7** Darwin's data

| Ordered observations $x_j$ ($j = 1, \ldots, 15$) | | | | | | | | | | | | | | |
|---|---|---|---|---|---|---|---|---|---|---|---|---|---|---|
| −67 | −48 | 6 | 8 | 14 | 16 | 23 | 24 | 28 | 29 | 41 | 49 | 56 | 60 | 75 |

Source: Box and Tiao (1968).

is the contour plot in the space of $\Delta = \mu_2 - \mu_1$ and $\pi_1$ of the profile likelihood obtained by maximizing the likelihood with respect to the other parameters. Note that in this example $\Delta$ denotes the Mahalanobis distance after multiplication by the standard deviation. The parameter space is restricted to $\Delta \geq 0$. Alternatively, $\Delta$ can be unrestricted with $\pi_1$ restricted to $\pi_1 \geq 0.5$. The dotted line in Figure 5 shows the "trough" of minima of the likelihood over $\pi_1$ for fixed $\Delta$. As pointed out by Aitkin and Rubin (1985), starting the EM algorithm with values of $(\Delta, \pi_1)$ on the "wrong" side of the trough leads to very slow convergence to a point of the boundary of the parameter space (corresponding to the solution for a single normal model) and not to the desired two-component mixture solution.

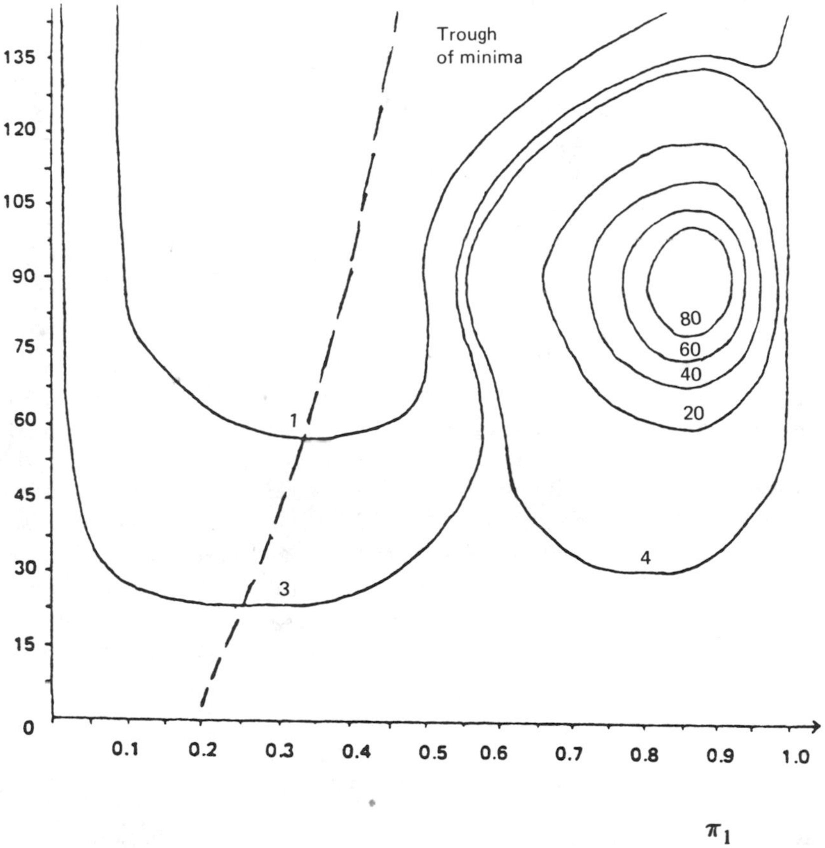


**FIGURE 5** Profile likelihood. (From Aitkin and Rubin, 1985.)

The increase in twice the log likelihood for a two-component normal mixture over the single normal model was found to be 6.9. This is significant at a 0.05 conventional level if, according to Wolfe's (1971) approximation, we take $-2\log\lambda$ to be $\chi_2^2$. The failure of the likelihood ratio test statistic $\lambda$ to have its usual asymptotic distribution under the null hypothesis when testing for the number of groups has been discussed at some length in Section 1.10. One alternative to using Wolfe's (1971) approximation for the null distribution of $-2\log\lambda$ is to adopt a bootstrap approach, as demonstrated in the previous section. Another proposal described in Section 1.10 is that of Aitkin and Rubin (1985) who proposed that inference in the mixture model may be based on

$$L(\boldsymbol{\theta}) = \log\left[\int \exp\{L(\boldsymbol{\pi},\boldsymbol{\theta})\}\, f_{\pi}(\boldsymbol{\pi})\, d\boldsymbol{\pi}\right],$$

where the vector of mixing proportions $\boldsymbol{\pi}$ is integrated out with respect to some prior distribution $f_{\pi}(\boldsymbol{\pi})$ placed on $\pi$. As a consequence, any null hypothesis about the number of groups is specified in the interior of the parameter space of $\boldsymbol{\theta}$, although $-2\log\lambda$ still may not have its usual asymptotic null distribution, as discussed in Section 1.10.

Aitkin and Rubin (1985) applied their method to Darwin's data with the prior distribution on $\pi_1$ taken to be a Beta distribution with

$$f_{\pi}(\pi_1) \propto (\pi_1 - \tfrac{1}{2})^{q_1}(1-\pi_1)^{q_2}, \qquad \tfrac{1}{2} \le \pi_1 < 1.$$

Let $\tilde{\boldsymbol{\theta}}$ denote the likelihood estimate of $\boldsymbol{\theta}$ using $L(\boldsymbol{\theta})$. Twice the difference between $L(\tilde{\boldsymbol{\theta}})$ and the log likelihood for the one sample normal model (that is, $-2\log\lambda$) is given for various selections of $q_1$ and $q_2$ in Table 8, which has been taken from Aitkin and Rubin (1985). Because of the small sample size the retention of the null hypothesis $H_0$ of one population (that is, no outliers) depends on the prior information. As summarized by Aitkin and Rubin (1985), if *a priori* outliers are very unlikely ($q_1$ large, $q_2$ small), the evidence in the data is discounted to the extent that $-2\log\lambda$ is not significant at the 0.05 level, taking $-2\log\lambda \sim \chi_1^2$ under $H_0$. On the other hand if *a priori* outliers are likely, then the data clearly indicate their presence.

## 3.4 CLUSTERING OF RARE EVENTS

Symons, Grimson and Yuan (1983) considered the clustering of cases of a rare disease, where the number of events observed for each unit is assumed to have a Poisson distribution with mean depending on the size of the unit

**TABLE 8** Value of $-2 \log \lambda$ for $B(q_1, q_2)$ prior on $\pi_1$

| | | $q_1$ | | | | | |
|---|---|---|---|---|---|---|---|
| | | −0.5 | 0 | 0.5 | 1 | 1.5 | 2 |
| | −0.5 | 4.8 | 4.5 | 4.1 | 3.6 | 3.1 | 2.7 |
| | 0 | 5.2 | 5.2 | 4.9 | 4.5 | 4.1 | 3.7 |
| | 0.5 | 5.3 | 5.5 | 5.4 | 5.1 | 4.8 | 4.5 |
| $q_2$ | | | | | | | |
| | 1 | 5.3 | 5.7 | 5.7 | 5.5 | 5.3 | 5.0 |
| | 1.5 | 5.3 | 5.8 | 5.9 | 5.8 | 5.6 | 5.4 |
| | 2 | 5.2 | 5.8 | 6.0 | 6.0 | 5.8 | 5.6 |

$\chi^2_{1,0.99} = 6.64$ $\chi^2_{1,0.95} = 3.84$ $\chi^2_{1,0.9} = 2.71$
Source: Aitkin and Rubin (1985).

and the cluster membership of that unit. They proposed three clustering criteria, including the mixture likelihood approach, and compared them in their application to the problem of identifying the counties at high risk to sudden infant death syndrome (SIDS) in North Carolina.

Data were available on the spatial occurrence of sudden infant deaths (SIDs) in $n = 100$ North Carolina counties over a four year period. The counties were presumed to be a mixture in proportions $\pi_1$ and $\pi_2$ of two groups $G_1$ and $G_2$, corresponding to normal and high risk with respect to SIDS. The number of events $x_j$ in the $j$th county was assumed to have a Poisson distribution with a rate $\mu_i$ in $G_i$ $(i = 1, 2; j = 1, \ldots, n)$. Let $\phi = (\pi_1, \pi_2, \mu_1, \mu_2)'$ and $\boldsymbol{\theta}_j = (\mu_1 M_j, \mu_2 M_j)'$ for $j = 1, \ldots, n$, where $M_j = N_j t_j$ was the mass of the population at risk in the $j$th county since there were $N_j$ individuals at risk over time $t_j$ $(j = 1, \ldots, n)$. The log likelihood under the mixture model is given by

$$L(\phi) = \sum_{j=1}^{n} \log \sum_{i=1}^{2} \pi_i f_i(x_j; \boldsymbol{\theta}_j),$$

where

$$f_i(x_j; \boldsymbol{\theta}_j) = \exp(-\mu_i M_j)(\mu_i M_j)^{x_j}/x_j! \qquad (i = 1, 2; j = 1, \ldots, n).$$

The EM algorithm can be easily applied to fit a mixture of $g$ Poisson distributions. The log likelihood for the complete data specification of the problem is

$$L_C(\phi) = \sum_{i=1}^{g}\sum_{j=1}^{h} z_{ij}\left\{\log \pi_i + x_j \log(\mu_i M_j) - \mu_i M_j\right\} - \sum_{j=1}^{n} \log(x_j!)$$

where, as usual, $z_{ij} = 1$ or 0 according as $x_j$ belongs to $G_i$ or not. It follows that the maximum likelihood estimates of $\pi_i$ and $\mu_i$ satisfy

$$\hat{\pi}_i = \sum_{j=1}^{n} \hat{\tau}_{ij}/n \tag{3.4.1}$$

and

$$\hat{\mu}_i = \sum_{j=1}^{n} \hat{\tau}_{ij} x_j \bigg/ \sum_{j=1}^{n} \hat{\tau}_{ij} M_j, \tag{3.4.2}$$

where

$$\hat{\tau}_{ij} = \hat{\pi}_i f_i(x_j; \hat{\theta}_j) \bigg/ \left\{\sum_{t=1}^{2} \hat{\pi}_t f_t(x_j; \hat{\theta}_j)\right\}$$

is the estimated posterior probability that $x_j$ belongs to $G_i$ $(i = 1, \ldots, g)$.

The equations (3.4.1) and (3.4.2) are solved iteratively, starting with some initial values for the $\hat{\tau}_{ij}$, and then computing the $\hat{\pi}_i$ and $\hat{\mu}_i$, which in turn are used to update the estimates of the $\tau_{ij}$, and so on, corresponding to alternation of the E and M steps until convergence.

Symons, Grimson and Yuan (1983) applied two other methods of clustering to this data set. One of these was the so-called classification likelihood approach described in general in Section 1.12. With this approach $\phi$ and $\mathbf{Z} = (\mathbf{z}_1', \ldots, \mathbf{z}_n')'$ are chosen to maximize $L_C(\phi)$; that is, the unobservable vector $\mathbf{z}_j$ of indicator variables associated with each $x_j$ is treated as an unknown parameter. The other method used was a Bayesian version of this last approach, whereby the clustering corresponds to the partition $\mathbf{Z}$ which maximizes

$$\int \exp\{L_C(\phi)\}\, f_\phi(\phi)\, d\phi,$$

using the vague prior

$$f_\phi(\phi) \propto \left(\prod_{i=1}^{g} \pi_i^{-1}\right)\left(\prod_{i=1}^{g} \mu_i^{-1}\right).$$

The results of partitioning the 100 counties of North Carolina into normal and high risk groups to SIDS as obtained by Symons, Grimson and Yuan (1983) are given in Table 9 for each of the three clustering methods used. The 20 counties with the largest SIDS death rates per 1000 live births are listed in decreasing order of that rate. The other 80 counties were put into the normal risk group (designated Lo in Table 9) by all the methods. It can be seen that the classification likelihood approach and the Bayesian version produced the same clusters, which differ for ten counties with the outright assignment on the basis of the estimated posterior probabilities of group membership under the mixture approach. As pointed out by Symons, Grimson and Yuan (1983), an important feature of the methods employed is that the clustering into normal and high risk groups is based not only on the SIDS rate of a county, but also on the number of live births. The variance of the estimated rate in a Poisson process decreases with increasing mass of the population at risk, and so a county with a high rate based upon a few births is not necessarily put in the high risk group.

The likelihood ratio test statistic for the test of the null hypothesis of $g = 1$ versus the alternative of $g = 2$ groups was computed by Symons, Grimson and Yuan (1982). The value of $-2\log\lambda$ was 61.15, which exceeded the largest of the 2,000 simulated values of $-2\log\lambda$, 16.269, suggesting a $P$-value of much less that 0.0005. The simulated values of $-2\log\lambda$ were formed from data generated according to a Poisson distribution with a common rate of 1/500 for each county, which is the maximum likelihood estimate from the original data, under the null hypothesis of a single rate for the counties. Hence their resampling approach is equivalent to the bootstrapping of the likelihood ratio test statistic described in Section 1.10. Symons, Grimson and Yuan (1983) concluded from their simulations that it would appear reasonable to approximate the upper tail of the null distribution of $-2\log\lambda$ by the $\chi_2^2$ distribution, although the latter provides a poor fit in the lower tail since a large number of zeros were observed for $-2\log\lambda$. It should be noted that the approximation of Wolfe (1971) applied here would give the $\chi_2^2$ distribution for $-2\log\lambda$ under the null hypothesis.

**TABLE 9** Results* for three versions of the likelihood approach to the clustering of 20 North Carolina counties with the largest SIDS death rate, 1974–1978

| County | No. of live births | No. of SIDs | Rate per 1000 | Rank | Method of Clustering: Classification | Bayes | Mixture |
|---|---|---|---|---|---|---|---|
| Anson | 1570 | 15 | 9.60 | 1 | Hi | Hi | Hi |
| Northampton | 1421 | 9 | 6.30 | 2 | Hi | Hi | Hi |
| Washington | 990 | 5 | 5.10 | 3 | Lo | Lo | Hi |
| Halifax | 3608 | 18 | 5.00 | 4 | Hi | Hi | Hi |
| Hertford | 1452 | 7 | 4.80 | 5 | Lo | Lo | Hi |
| Hoke | 1494 | 7 | 4.70 | 6 | Lo | Lo | Hi |
| Greene | 870 | 4 | 4.60 | 7 | Lo | Lo | Hi |
| Bertie | 1324 | 6 | 4.53 | 8 | Lo | Lo | Hi |
| Bladen | 1782 | 8 | 4.49 | 9 | Lo | Lo | Hi |
| Columbus | 3350 | 15 | 4.48 | 10 | Hi | Hi | Hi |
| Swaine | 675 | 3 | 4.40 | 11 | Lo | Lo | Lo |
| Warren | 968 | 4 | 4.10 | 12 | Lo | Lo | Lo |
| Rutherford | 2992 | 12 | 4.00 | 13 | Lo | Lo | Hi |
| Robeson | 7889 | 31 | 3.90 | 14 | Hi | Hi | Hi |
| Lincoln | 2216 | 8 | 3.61 | 15 | Lo | Lo | Hi |
| Rockingham | 4449 | 16 | 3.60 | 16 | Lo | Lo | Hi |
| Scotland | 2255 | 8 | 3.50 | 17 | Lo | Lo | Hi |
| Pender | 1228 | 4 | 3.30 | 18 | Lo | Lo | Lo |
| Wilson | 3702 | 11 | 3.00 | 19 | Lo | Lo | Lo |
| Lenoir | 3589 | 10 | 2.90 | 20 | Lo | Lo | Lo |

*The other 80 of the 100 North Carolina counties were clustered as "Lo" risk by all three methods.

Source: Adapted from M.J. Symons, R.C. Grimson and Y.C. Yuan (1983), Clustering of rare events. *Biometrics* **39**, 193–205. With permission from the Biometric Society.

## 3.5 LATENT CLASSES OF TEACHING STYLES

As part of their detailed statistical modeling of an extensive body of educational research data on teaching, Aitkin, Anderson and Hinde (1981) fitted latent class models to data on 468 teachers measured on 38 binary variables describing teaching behavior. The data were collected from a teacher questionnaire containing 28 items which were coded into 38 binary items, labeled 1 to 38 here. Accordingly, the observation vector $\mathbf{x}$ has $p = 38$ dimensions, with its $v$th element recorded as zero or one corresponding to a negative or positive response to the $v$th item on the questionnaire $(v = 1, \ldots, p)$.

Under the mixture approach adopted by Aitkin, Anderson and Hinde (1981) there are $g$ latent (that is, unobservable) classes $G_1, \ldots, G_g$ in proportions $\pi_1, \ldots, \pi_g$, underlying the different teaching styles, so that the probability function of $\mathbf{x}$ is

$$f(\mathbf{x}; \phi) = \sum_{i=1}^{g} \pi_i f_i(\mathbf{x}; \theta),$$

where $f_i(\mathbf{x}; \theta)$ is the probability function of $\mathbf{x}$ in $G_i$ $(i = 1, \ldots, g)$ and $\phi = (\pi', \theta')'$. Again we let $\boldsymbol{\mu}_i$ denote the mean of $\mathbf{x}$ in $G_i$, and we let $\mu_{iv}$ and $x_{jv}$ denote the $v$th element of $\boldsymbol{\mu}_i$ and of the $j$th observation $\mathbf{x}_j$, respectively. Here for each $\mathbf{x}_j$,

$$\mu_{iv} = \mathrm{pr}\{x_{jv} = 1 \mid \mathbf{x}_j \in G_i\} \qquad (i = 1, \ldots, g; v = 1, \ldots, p).$$

Following the approach of Lazarsfeld and Henry (1968), which has been widely used in latent class modeling, Aitkin, Anderson and Hinde (1981) assumed conditional independence so that

$$f_i(\mathbf{x}_j; \theta) = \prod_{v=1}^{p} \mu_{iv}^{x_{jv}} (1 - \mu_{iv})^{1 - x_{jv}},$$

where the vector $\theta$ of unknown parameters in the class probability functions consists now of only the class means. The assumption of independence within a group implies that the correlations between item responses are due solely to the group differences in the teaching styles.

The maximum likelihood estimate of $\phi$ can be easily computed using the EM algorithm, and the details have been given by Aitkin (1980b); see also Everitt (1984b). The estimates of $\pi_i$ and $\mu_{iv}$ $(i = 1, \ldots, g; v = 1, \ldots p)$ so obtained for $g = 2$ and 3 latent classes are displayed in Table 10, while the value of the likelihood ratio test statistic $-2 \log \lambda$ for testing $g$ against $g + 1$ classes is listed in Table 11. The results in these tables have been taken from Aitkin, Anderson and Hinde (1981).

As discussed in Section 1.10, the aforementioned authors adopted a bootstrap approach to the difficult problem of assessing the number of components in a mixture. This approach, which was applied in the example just considered in Section 3.2, was used in their analysis of the present data set to assess the statistical significance of the clusters produced. In the test of $g = 1$ versus $g = 2$ classes, the value of 775.8 for $-2 \log \lambda$ formed from the

**TABLE 10** Maximum likelihood estimates expressed as percentages for $g = 2$ and 3 latent classes for teacher data

| | | $g = 2$ | | $g = 3$ | | |
|---|---|---|---|---|---|---|
| $v$ | Item | $\hat{\mu}_{1v}$ | $\hat{\mu}_{2v}$ | $\hat{\mu}_{1v}$ | $\hat{\mu}_{2v}$ | $\hat{\mu}_{3v}$ |
| 1 | Pupils have choice in where to sit | 22 | 43 | 20 | 44 | 33 |
| 2 | Pupils sit in groups of three or more | 60 | 87 | 54 | 88 | 79 |
| 3 | Pupils allocated to seating by ability | 35 | 23 | 36 | 22 | 30 |
| 4 | Pupils stay in same seats for most of day | 91 | 63 | 91 | 52 | 89 |
| 5 | Pupils not allowed freedom of movement in classroom | 97 | 54 | 100 | 53 | 74 |
| 6 | Pupils not allowed to talk freely | 89 | 48 | 94 | 50 | 61 |
| 7 | Pupils expected to ask permission to leave room | 97 | 76 | 96 | 69 | 95 |
| 8 | Pupils expected to be quiet | 82 | 42 | 92 | 39 | 56 |
| 9 | Monitors appointed for special jobs | 85 | 67 | 90 | 70 | 69 |
| 10 | Pupils taken out of school regularly | 32 | 60 | 33 | 70 | 35 |
| 11 | Timetable used for organizing work | 90 | 66 | 95 | 62 | 77 |
| 12 | Use own materials rather than textbooks | 19 | 49 | 20 | 56 | 26 |
| 13 | Pupils expected to know tables by heart | 92 | 76 | 97 | 80 | 75 |
| 14 | Pupils asked to find own reference materials | 29 | 37 | 28 | 39 | 34 |
| 15 | Pupils given homework regularly | 35 | 22 | 45 | 29 | 12 |
| 16 | Teacher talks to whole class | 71 | 44 | 73 | 37 | 62 |
| 17 | Pupils work in groups on teacher tasks | 29 | 42 | 24 | 45 | 38 |
| 18 | Pupils work in groups on work of own choice | 15 | 46 | 13 | 59 | 20 |
| 19 | Pupils work individually on teacher tasks | 55 | 37 | 57 | 32 | 50 |
| 20 | Pupils work individually on work of own choice | 28 | 50 | 29 | 60 | 26 |
| 21 | Explore concepts in number work | 18 | 55 | 14 | 62 | 34 |
| 22 | Encourage fluency in written English even if inaccurate | 87 | 94 | 87 | 95 | 90 |
| 23 | Pupils work marked or graded | 43 | 14 | 50 | 16 | 20 |
| 24 | Spelling and grammatical errors corrected | 84 | 68 | 86 | 64 | 78 |
| 25 | Stars given to pupils who produce best work | 57 | 29 | 65 | 30 | 34 |
| 26 | Arithmetic tests given at least once a week | 59 | 38 | 68 | 43 | 35 |
| 27 | Spelling tests given at least once a week | 73 | 51 | 83 | 56 | 46 |
| 28 | End of term tests given | 66 | 44 | 75 | 48 | 42 |
| 29 | Many pupils who create discipline problems | 9 | 9 | 7 | 1 | 18 |
| 30 | Verbal reproof sufficient | 97 | 95 | 98 | 99 | 91 |
| 31 | Discipline—extra work given | 70 | 53 | 69 | 49 | 67 |
| 32 | Discipline—smack | 65 | 42 | 64 | 33 | 63 |
| 33 | Discipline—withdrawal of privileges | 86 | 77 | 85 | 74 | 85 |
| 34 | Discipline—send to head teacher | 24 | 17 | 21 | 13 | 28 |

**TABLE 10** *Continued*

| | | $g=2$ | | $g=3$ | | |
|---|---|---|---|---|---|---|
| $v$ | Item | $\hat{\mu}_{1v}$ | $\hat{\mu}_{2v}$ | $\hat{\mu}_{1v}$ | $\hat{\mu}_{2v}$ | $\hat{\mu}_{3v}$ |
| 35 | Discipline—Send out of room | 19 | 15 | 15 | 8 | 27 |
| 36 | Emphasis on separate subject teaching | 85 | 50 | 87 | 43 | 73 |
| 37 | Emphasis on aesthetic subject teaching | 55 | 63 | 53 | 61 | 63 |
| 38 | Emphasis on integrated subject teaching | 22 | 65 | 21 | 75 | 33 |
| ... | | | | | | |
| | Estimated proportion of teachers in each class | 0.54 | 0.46 | 0.37 | 0.31 | 0.32 |

Adapted from Aitkin, Anderson and Hinde (1981).

original data well exceeded 84.4, the largest of the 19 bootstrap values of $-2\log\lambda$ generated, so the null hypothesis of a homogeneous population for teaching style was formally rejected at the 0.05 level. In the test of $g=2$ versus $g=3$ classes, the null hypothesis was again rejected at the 0.05 level, with the observed value of $-2\log\lambda$, 184.7, being greater than the largest (87.8) of the 19 bootstrap values generated for this second test. Although the likelihood ratio test statistic was assessed as quite large relative to the critical value for both the homogeneity and two-class null hypotheses, Aitkin, Anderson and Hinde (1981) did not use or interpret models with $g>3$ classes because of the difficulty with multiple maxima. The existence of the latter is not surprising, since each extra class requires an additional

**TABLE 11** Likelihood ratio test statistic for $g$ versus $g+1$ latent classes of teaching styles

| $g$ | $-2\log\lambda$ |
|---|---|
| 1 | 775.8 |
| 2 | 184.7 |
| 3 | 173.8 |
| 4 | 142.5 |
| 5 | 126.0 |
| 6 | 121.9 |
| 7 | 96.3 |

set of 39 parameters to be estimated. Only one solution of the likelihood equation was found for $g = 2$ classes. However, for $g = 3$, they located another solution besides the one given in Table 10.

They also investigated the number of latent classes by an informal graphical procedure, which also provided some evidence of the fit of the latent classes with the assumption of conditional independence. The procedure was based on

$$\bar{x}_j = \sum_{v=1}^{p} x_{jv}/p \qquad (j = 1, \ldots, n)$$

where $p\bar{x}_j$ was referred to as the total formality score. The $p = 38$ items were relabeled so that $x_{jv} = 1$ represents the "formal" and $x_{jv} = 0$ the "informal" end of the range for $v = 1, \ldots, p$. Under the null hypothesis of $g = 1, \bar{x}_j$ is approximately normal while if there are $g$ latent classes and the conditional independence model holds, $\bar{x}_j$ is distributed approximately as a normal mixture with $g$ components in proportions $\pi_1, \ldots, \pi_g$. A single normal and a mixture of $g = 2$ and 3 normal densities were fitted in turn to the distribution of the total formality scores for the 468 teachers. The chi-squared goodness-of-fit test provided strong evidence in favor of three overlapping rather than two distinct, latent classes. The fit of the former model in the body of the distribution was good.

Concerning the interpretation of the teaching styles under the latent class model, Aitkin, Anderson and Hinde (1981) noted that the first latent class $G_1$ is at the formal end of every item in Table 10 in the $g = 2$ model, and in the $g = 3$ model except for items 29, 34, and 35. The second class $G_2$ is at the informal end of every item in the $g = 2$ model, and in the $g = 3$ model apart from items 13, 15, 26, 27, 28, and 37 . In the latter model, the third class is intermediate between $G_1$ and $G_2$. The reader is referred to the actual article by Aitkin, Anderson and Hinde (1981) for a detailed account of the results in Table 10. More recently, Aitkin and Rubin (1985) also considered this data set with their method whereby a prior distribution is placed on the vector of mixing proportions. Four prior distributions were examined and, as they anticipated, the data overwhelmed any realistic prior assumption.

# 4
# Estimation of Mixing Proportions

## 4.1 INTRODUCTION

In this chapter, we consider the problem where the superpopulation $G$ is a genuine mixture of $g$ distinct populations $G_1, \ldots, G_g$, and the primary aim is to estimate the proportions $\pi_1, \ldots, \pi_g$ in which they occur. This problem has attracted much attention in practice. Frequently, in addition to the unclassified sample, there are also available some data of known origin from each population. Usually the latter have been obtained by sampling separately from each population, and so provide no information about the mixing proportions.

Some examples where the above model is appropriate have been given by Hosmer (1973b) in the context of identifying the sex of halibut and by Odell (1976) and Odell and Basu (1976) on the estimation of crop acreages from remote sensors on orbiting satellites. The last two references have been taken from the special issue of *Communications in Statistics* on Remote Sensing (1976, **A5**, 1077-1191), which is a source of additional references on this problem. In the former example, the aim is to estimate the proportion of each sex of halibut on the basis of their length; their sex, unlike their age and length, cannot be identified by humans from external characteristics. For various age classes, data on the lengths of halibut are available from research cruises where the sex of each fish is identified.

In the example of crop acreage estimation, remotely sensed observations are available from a mixture of several crops. The problem is to estimate the acreage of a particular crop as a proportion of the total acreage. Data are often available on each of the crops to provide estimates of the unknown parameters in the distribution of an individual crop. Switzer (1980) used remotely sensed data to estimate the percentage of component terrains within a desired surface area. As in the previous example, the process is based on digitized imagery (from the Landsat earth satellite) which consists of an integrated energy measurement for each of four wavelength bands reported for surface area elements, pixels, of about an acre in size. The distribution of this data vector depends on the physical properties of the surface, and so one can attempt to identify the surface type corresponding to each pixel. There is available a sample of "training" pixels of known surface type for each of the contemplated categories to which pixels might be assigned. There is now a vast literature on these problems in remote sensing. For example, Heydorn (1984) has surveyed various estimators used to estimate crop proportions from remotely sensed measurements.

Another example is concerned with digitized images of cervical cells from the PAP smear slide where one has to assess the proportions in which cells of various types occur. This is an important step in the diagnosis of a cervical cytology specimen (White and Castleman, 1981). A botanical example may be found in Gordon (1982), who noted that for pollen types whose morphological characteristics overlap markedly, it is often useful to be able to estimate the proportions of each type present in a sample of pollen grains. An example where the components of the mixture are not taken to belong to the same parametric family was given by Brownie, Habicht and Robson (1983), who were concerned with the estimation of the prevalence of anemia from a sample of hemoglobin values. Other examples are to be presented shortly in Sections 4.7 and 4.8.

It has been explained in Chapters 1 and 2 how a mixture model can be fitted to data, and how the unknown parameters, including the mixing proportions, can be estimated according to the likelihood approach. Some additional references in this context to those given previously are Laird (1978), who considered nonparametric likelihood estimation of a mixing distribution, and Lindsay (1983) who linked his general theory, developed from a study of the geometry of the mixture likelihood, to the problem of estimating a discrete mixing distribution. An account of the Bayesian solution to this problem, where the observations are obtained sequentially, has been given by Titterington, Smith and Makov (1985, Chapter 6).

As mentioned at the beginning of this chapter, if there are data of known origin available, they usually have been obtained by sampling separately from each of the component populations, and so provide no information about the unknown mixing proportions. Therefore, in seeking a suitable starting value for $\pi$ in the application of the EM algorithm to this problem, we are led to a consideration of some easily computed estimates of $\boldsymbol{\pi}$ for initial values. Various estimates of the mixing proportions which are useful in this respect are to be discussed in the following sections. A case study is to be described in Section 4.7, while in the last section of the chapter, an example is given involving the construction of a test for the homogeneity of mixing proportions for different mixtures of the same component distributions.

## 4.2 LIKELIHOOD ESTIMATION

We have seen from Section 1.6, that in fitting a mixture of $g$ components to a completely unclassified sample $\mathbf{x}_1, \ldots, \mathbf{x}_n$, the likelihood estimate of $\pi_i$ is given by

$$\hat{\pi}_i = \sum_{j=1}^{n} \hat{\tau}_{ij} \qquad (i = 1, \ldots, g), \tag{4.2.1}$$

where the $\hat{\tau}_{ij}$ are the estimated posterior probabilities of component membership for $\mathbf{x}_j$ $(j = 1, \ldots, n)$, formed by replacing $\phi$ with its likelihood estimate $\hat{\phi}$ in (1.4.3). The $\hat{\pi}_i$ are computed iteratively along with the estimate of $\boldsymbol{\theta}$ containing the unknown parameters in the component distributions. This iterative process corresponding to an application of the EM algorithm has been described in Section 1.6 for arbitrary component distributions, and in Section 2.1 for normal components. We shall be concentrating on this particular case, where

$$\mathbf{x} \sim N(\boldsymbol{\mu}_i, \boldsymbol{\Sigma}_i) \qquad \text{in } G_i \text{ with prob. } \pi_i \quad (i = 1, \ldots, g), \tag{4.2.2}$$

for the remainder of the chapter.

In the event of there being some classified data available, we adhere to our previous notation and let $\mathbf{y}_{ij}$ $(j = 1, \ldots, m_i)$ be the $m_i$ observations of known origin from $G_i$ $(i = 1, \ldots, g)$, and $m = \sum_{i=1}^{i=g} m_i$. Hosmer (1973a) and Hosmer and Dick (1977) have discussed various models under which the $m$ classified observations may have been obtained. If the classified data were obtained by sampling from the mixture $G$, then an initial estimate of $\pi_i$ is available through $m_i/m$ $(i = 1, \ldots, g)$. Of course, if $m$ were very large relative to $n$, then there may be no need to consider undertaking the estimation of $\pi_i$ on the basis of the full set of $m + n$ observations. But in the typical situation in practice, $m$ is small relative to $n$, as procuring observations of known origin is generally more difficult, or at least more expensive, than obtaining unclassified data.

The likelihood estimate of $\pi_i$ on the basis of the combined data, where the classified data were sampled from the mixture $G$, is given by

$$\hat{\pi}_i = \left( m_i + \sum_{j=1}^{n} \hat{\tau}_{ij} \right) \Big/ (m+n) \qquad (i = 1, \ldots, g)$$

and, under (4.2.2), the $\hat{\boldsymbol{\mu}}_i$ and the $\widehat{\boldsymbol{\Sigma}}_i$ are computed according to (2.1.8) and (2.1.9). As noted in Section 1.11, the likelihood will be bounded if there are at least $p+1$ classified observations from each normal component population.

If the classified data provide no information on the mixing proportions, for example, having been obtained by sampling separately from each of the populations, then the likelihood estimate of $\pi_i$ is given by (4.2.1), but the $\hat{\boldsymbol{\mu}}_i$ and the $\widehat{\boldsymbol{\Sigma}}_i$ are still computed from the combined data according to (2.1.8) and (2.1.9). In this case $m_i/m$ cannot be considered as an estimate of $\pi_i$, and so a suitable starting value for $\pi_i$ in the iterative computation of the likelihood solution must be found. Some easily computed estimates of the mixing proportions, useful in this respect, are discussed now.

## 4.3 DISCRIMINANT ANALYSIS ESTIMATOR

In the case where there are available data of known origin from each component population an obvious and computationally straightforward way of proceeding is to form a discriminant rule $R$ from these classified data, and then to apply it to the unclassified data $\mathbf{x}_1, \ldots, \mathbf{x}_n$ to find the proportion of these $n$ observations assigned to the $i$th population $G_i$ $(i = 1, \ldots, g)$. That is, if $\tilde{n}_i$ denotes the number of the $n$ unclassified observations assigned to $G_i$ by $R$, then a rough estimate of $\pi_i$ is provided by $\tilde{n}_i/n$. We shall reserve the notation $n_i$ for the number of the unclassified observations actually from $G_i$.

In order to correct this estimate $\tilde{n}_i/n$ for bias, we need to know the conditional allocation rates of $R$. We let $e_{ij}(\phi;\ \mathbf{Y})$ be the probability, conditional on

$$\mathbf{Y} = \left\{ \mathbf{y}_{ij}, i = 1, \ldots, g; j = 1, \ldots, m_i \right\},$$

that a randomly chosen member of $G_i$ is allocated to $G_j$ by $R$ $(i, j = 1,$

..., $g$), where

$$\sum_{j=1}^{g} e_{ij}(\phi; \mathbf{Y}) = 1 \qquad (i = 1, \ldots, g).$$

If we view the rule $R$ as dividing the sample space into $g$ disjoint regions $C_1, \ldots, C_g$ in which $\mathbf{x}$ is assigned to $G_i$ if it falls in $C_i$ ($i = 1, \ldots, g$), then we can write

$$e_{ij}(\phi; \mathbf{Y}) = \mathrm{pr}\{\mathbf{x} \in C_j \mid \mathbf{x} \in G_i; \mathbf{Y}\} \qquad (i, j = 1, \ldots, g).$$

For $g = 2$, it can be easily seen that

$$E(\tilde{n}_1/n) = \pi_1 e_{11} + \pi_2 e_{21}, \tag{4.3.1}$$

and

$$E(\tilde{n}_2/n) = \pi_1 e_{12} + \pi_2 e_{22}, \tag{4.3.2}$$

with either equation giving the so-called discriminant analysis estimator of $\pi_1$,

$$\hat{\pi}_{1D} = (\tilde{n}_1/n - e_{21})/(e_{11} - e_{21}) \tag{4.3.3}$$

as an unbiased estimator of $\pi_1$. In the above equations $e_{ij}(\phi; \mathbf{Y})$ is written simply as $e_{ij}$ for convenience. If $\hat{\pi}_{1D}$ is outside [0,1], then it is assigned the appropriate value zero or one.

On considering (4.3.1) and (4.3.2) simultaneously, $\hat{\pi}_{1D}$ and $\hat{\pi}_{2D} = 1 - \hat{\pi}_{1D}$ can be expressed equivalently as

$$\hat{\boldsymbol{\pi}}_D = \mathbf{J}^{-1}(\tilde{n}_1/n, \tilde{n}_2/n)', \tag{4.3.4}$$

where $\hat{\boldsymbol{\pi}}_D = (\hat{\pi}_{1D}, \hat{\pi}_{2D})'$ and

$$\mathbf{J} = \begin{bmatrix} e_{11} & e_{21} \\ e_{12} & e_{22} \end{bmatrix}.$$

Odell and Basu (1976) remarked that $\mathbf{J}$ has been called the confusion matrix. Hence $\hat{\boldsymbol{\pi}}_D$ is sometimes referred to as the confusion matrix estimator rather than the discriminant analysis estimator.

The expression (4.3.4) can be generalized to $g > 2$ populations to give

$$\hat{\boldsymbol{\pi}}_D = \mathbf{J}^{-1}(\tilde{n}_1/n, \ldots, \tilde{n}_g/n)', \tag{4.3.5}$$

where the $(i, j)$th element of $\mathbf{J}$ is equal to

$$(\mathbf{J})_{ij} = e_{ji} \qquad (i, j = 1, \ldots g).$$

Note that for known $e_{ij}$, $\hat{\boldsymbol{\pi}}_D$ is the maximum likelihood estimate of $\boldsymbol{\pi}$ based on the proportions $\tilde{n}_i/n$ $(i = 1, \ldots, g)$, and is unbiased. According to Macdonald (1975), $\hat{\boldsymbol{\pi}}_D$ seems to have been first suggested by Worlund and Fredin (1962). For ungrouped data from a univariate mixture with known component distributions, Boes (1966) derived a class of estimators of the mixing proportions from consideration of the attainment of the Cramér-Rao bound. His $\boldsymbol{\pi}^0$-efficient estimator was unbiased and fully efficient when $\boldsymbol{\pi} = \boldsymbol{\pi}^0$. For $g = 2$, he showed that his $\boldsymbol{\pi}^0$-efficient unbiased estimator can be obtained by using (4.3.3), where $C_1 = (-\infty, x)$ and $C_2$ is the complement of $C_1$, and then taking a suitable weighted average of the $x$ values. Pau and Chen (1977) and Kittler and Pau (1978) developed essentially the same estimator as (4.3.3) in the context of a pattern recognition system for quality control on the basis of lot acceptance sampling. Other related references include Guseman and Walton (1977, 1978).

For grouped data with the number of regions or categories $C_j$, say $N$, not necessarily equal to the number of components $g$ in the mixture, Macdonald (1975) described a class of weighted least squares estimators based on the observed proportions $n_j/n$ $(j = 1, \ldots, N)$. They were obtained by minimizing the weighted squared differences between $\tilde{n}_j/n$ and its expected value (the $j$th category probability)

$$\sum_{i=1}^{g} \pi_i e_{ij}$$

over $j = 1, \ldots, N$. The discriminant analysis estimator can be considered a member of this class since it corresponds to $N = g$ and to unit weights.

James (1978) studied $\hat{\boldsymbol{\pi}}_D$ for a mixture of two univariate normal populations with known means and variances (not necessarily equal) and with $R(x) \equiv x$. An unclassified observation $x_j$ was assigned to one of two regions according as it was greater or less than a fixed cut-off point $k$. It was noted that one reasonable choice of $k$ is the value which maximizes the sum of the correct allocation rates of $R$, which is $\frac{1}{2}(\mu_1 + \mu_2)$, as suggested also by Johnson (1973). James (1978) extended his approach to the case of three regions through the use of two cut-off points $k_1$ and $k_2$ and, as in the case of two regions, investigated the selection of the cut-off points so as to minimize the variance of the resulting estimator of $\pi_1$. It was emphasized that these estimators do not require precise measurement of the observations as only their size relative to the cut-off points need be recorded.

Concerning the choice of the rule $R$ in the formation of the discriminant analysis estimator $\hat{\boldsymbol{\pi}}_D$ for a mixture of an arbitrary number of multivariate component populations, we might simply base $R$ on the Mahalanobis distance, where an unclassified observation $\mathbf{x}_j$ is allocated to $G_i$ if

$$D(\mathbf{x}_j, \bar{\mathbf{y}}_i; \mathbf{S}_i) < D(\mathbf{x}_j, \bar{\mathbf{y}}_t; \mathbf{S}_t) \qquad (t = 1, \ldots, g; t \neq i), \tag{4.3.6}$$

where $\bar{\mathbf{y}}_i$ and $\mathbf{S}_i$ denote the sample mean and covariance matrix of the classified data from $G_i$, as given by (2.5.2) and (2.5.3) respectively ($i = 1, \ldots, g$). Another choice is to take $R$ to be the rule which assigns $\mathbf{x}_j$ to $G_i$ if

$$\hat{f}_i(\mathbf{x}_j) > \hat{f}_t(\mathbf{x}_j) \qquad (t = 1, \ldots, g; t \neq i). \tag{4.3.7}$$

For the estimative approach,

$$\hat{f}_i(\mathbf{x}) = f_i(\mathbf{x}; \hat{\boldsymbol{\theta}}),$$

where $\hat{\boldsymbol{\theta}}$ is the likelihood estimate of $\boldsymbol{\theta}$ based on the classified data. Under the normality assumption (4.2.2), $f_i(\mathbf{x}; \hat{\boldsymbol{\theta}})$ is the multivariate normal density with mean $\bar{\mathbf{y}}_i$ and, ignoring the difference of one in the divisor, covariance matrix $\mathbf{S}_i$ ($i = 1, \ldots, g$), and so (4.3.7) is equivalent to (4.3.6) if the pooled sample covariance matrix $\mathbf{S}$ is used in each component density. For the fully Bayesian approach (Aitchison and Dunsmore, 1975, Chapter 2) under (4.2.2), the $\hat{f}_i(\mathbf{x})$ are given by the predictive densities (2.6.9) or (2.6.7), depending on whether the normal components have the same covariance matrix or not.

Nonparametric estimates of the component densities may be used in (4.3.7), using the kernel method as considered, for example, by Habbema, Hermans and van den Broek (1974), Aitchison and Aitken (1976), Titterington (1980), and Hall (1981). Other references may be found in the monograph by Hand (1982) devoted entirely to the role of the kernel method in discriminant analysis. More recently, in the case where the classified data have been sampled from the mixture $G$, Hall and Titterington (1985) have given a very simple technique for combining the unclassified and classified data to produce a nonparametric estimator of $f_i(\mathbf{x}; \boldsymbol{\theta})$ with smaller variance than that of the estimator based solely on the classified data.

As cautioned by Hall (1981), the choice of the smoothing parameter or window size is crucial in the use of the Rosenblatt-Parzen class of kernel density estimators. Also, Hall (1986) has shown that a kernel with thin tails, such as the standard normal or the double exponential, is often not a good choice when using likelihood cross-validation to choose the smoothing parameter. A suitable univariate kernel suggested by Hall (1986) is

$$0.1438 \exp\left[-\tfrac{1}{2}\{\log(1+|x|)\}^2\right].$$

Other approaches to the construction of the rule $R$ include logistic regression as proposed initially by Cox (1966) and Day and Kerridge (1967). Whether the classified data were obtained under a separate or mixture sampling scheme is not crucial to the likelihood estimation of the logistic discriminant coefficients; see, for example, Anderson (1972) and McLachlan (1980a). Another approach is to base $R$ on the combinatoric procedure developed by Dunn and Smith (1980, 1982) and Dunn (1982).

Although the discriminant rule $R$ would usually be chosen so as not to depend on $\boldsymbol{\pi}$, its conditional allocation rates $e_{ij}$ will depend on the unknown parameters in the component distributions, and so must be estimated from the classified data. Regardless of the choice of $R$, the allocation rates $e_{ij}$ will generally have to be estimated nonparametrically, as parametric estimators are available only in special instances, for example, for $g = 2$ normal populations with the same covariance matrix. The conditional allocation rate $e_{ij}$ can be estimated by the proportion of the $m_i$ observations $\mathbf{y}_{ik}$ $(k = 1, \ldots, m_i)$ assigned to $G_j$, when $R$ is trained on the classified data $(i, j = 1, \ldots, g)$. It is well known that these apparent allocation rates provide too optimistic an assessment (see, for example, McLachlan, 1976), but they can be corrected for bias nonparametrically using either cross-validation (Lachenbruch and Mickey, 1968) or the bootstrap method (Efron, 1982 and 1983). Further references on these estimators may be found in McLachlan (1980b, 1986a, 1987).

It can be seen that if a nonparametric rule $R$ is adopted and the allocation rates are estimated nonparametrically, then the discriminant analysis estimator can be made distribution-free. In this sense, it should be more robust than the likelihood estimator in estimating the mixing proportions. Tubbs and Coberly (1976) have studied the robustness of several estimators of the mixing proportions when the component distributions are subjected to changes in location.

## 4.4 ASYMPTOTIC RELATIVE EFFICIENCY OF DISCRIMINANT ANALYSIS ESTIMATOR

It is of interest to consider the relative efficiency of the discriminant analysis estimator $\hat{\boldsymbol{\pi}}_D$. Ganesalingam and McLachlan (1981) derived the asymptotic efficiency of $\hat{\pi}_{1D}$ relative to $\hat{\pi}_1$ in the case of a mixture of two normal populations with a common covariance matrix, and where the available classified data provided no information on the mixing proportion. The asymptotic relative efficiency of $\hat{\pi}_{1D}$ was defined by the ratio of the coefficients of $1/(m+n)$ in the asymptotic variances of $\hat{\pi}_1$ and $\hat{\pi}_{1D}$ expanded in powers of $1/(m+n)$. Since the biases of $\hat{\pi}_1$ and $\hat{\pi}_{1D}$ are of the first order with respect to $1/(m+n)$, the mean squared errors of $\hat{\pi}_1$ and $\hat{\pi}_{1D}$ are equal to their variances, ignoring terms of the second order.

In deriving the asymptotic relative efficiency of $\hat{\pi}_{1D}$, the discriminant rule $R$ was based essentially on Fisher's linear discriminant function, where an observation $\mathbf{x}$ is allocated to $G_1$ or $G_2$ according as

$$\{\mathbf{x} - \tfrac{1}{2}(\bar{\mathbf{y}}_1 + \bar{\mathbf{y}}_2)\}' \mathbf{S}^{-1}(\bar{\mathbf{y}}_2 - \bar{\mathbf{y}}_1) \tag{4.4.1}$$

is less or greater than zero, where the $\bar{\mathbf{y}}_i$ and $\mathbf{S}$ denote the sample means and pooled sample covariance matrix formed from the classified data. The use of a zero cut-off point with (4.4.1) minimizes asymptotically the sum of the associated misallocation rates.

The conditional allocation rates associated with (4.4.1), $e_{12}$ and $e_{21}$, are obtained by putting $i = 1$ and 2 respectively in

$$\Phi\left((-1)^{i+1} \frac{\{\boldsymbol{\mu}_i - \frac{1}{2}(\bar{\mathbf{y}}_1 + \bar{\mathbf{y}}_2)\}' \mathbf{S}^{-1}(\bar{\mathbf{y}}_2 - \bar{\mathbf{y}}_1)}{\{(\bar{\mathbf{y}}_2 - \bar{\mathbf{y}}_1)'\mathbf{S}^{-1}\boldsymbol{\Sigma}\mathbf{S}^{-1}(\bar{\mathbf{y}}_2 - \bar{\mathbf{y}}_1)\}^{1/2}}\right), \tag{4.4.2}$$

where $\Phi$ denotes the standard normal distribution function. In their derivation of the asymptotic relative efficiency of $\hat{\pi}_{1D}$, Ganesalingam and McLachlan (1981) estimated (4.4.2) by replacing $\boldsymbol{\mu}_i$ and $\boldsymbol{\Sigma}$ with $\bar{\mathbf{y}}_i$ and $\mathbf{S}$ respectively to give

$$\hat{e}_{12} = \hat{e}_{21} = \Phi\left(-\tfrac{1}{2}\hat{\Delta}\right), \tag{4.4.3}$$

where

$$\hat{\Delta}^2 = (\bar{\mathbf{y}}_1 - \bar{\mathbf{y}}_2)'\mathbf{S}^{-1}(\bar{\mathbf{y}}_1 - \bar{\mathbf{y}}_2).$$

They noted that since the asymptotic variance of $\hat{\pi}_{1D}$ was to be considered up to and including terms of the first order only, the plug-in estimate (4.4.3) suffices, although in practice it may be modified by the addition of some first and second order terms so that the bias is of the third order only (McLachlan, 1974 and 1975b).

It was shown that the asymptotic relative efficiency depends on the parameters $\pi_1$, $\Delta$, $\gamma = n/(m+n)$, and $m_1/m$, but not explicitly on the number of dimensions $p$, as the asymptotic variances of $\hat{\pi}_1$ and $\hat{\pi}_{1D}$ do not depend on this. In Table 12, the asymptotic relative efficiency of $\hat{\pi}_{1D}$ is listed as a percentage for various combinations of the parameters $\pi_1$, $\Delta$, and $\gamma$ with $m_1 = m_2$. These values were extracted from Ganesalingam and McLachlan (1981) who noted that the asymptotic relative efficiency exhibited essentially the same characteristics for $m_1 \neq m_2$ as for $m_1 = m_2$. Since $m_1 = m_2$ in Table 12, interchanging $\pi_1$ with $1 - \pi_1$ does not alter the values of the asymptotic relative efficiency, and so the table provides results also for $\pi_1 > 0.5$.

Briefly, it can be seen from Table 12 for $\pi_1 \leq 0.25$ that there may be a substantial loss in efficiency if $\hat{\pi}_{1D}$ were used instead of $\hat{\pi}_1$. Indeed, for such mixing proportions there is generally a considerable loss in efficiency if the sample contains a high proportion of unclassified observations as evidenced for $\gamma \geq 0.75$. Hence if $\hat{\pi}_{1D}$ suggests that the mixing proportions are fairly disparate, then it is recommended that one should proceed further and compute the asymptotically efficient estimator $\hat{\pi}_1$, particularly if the sample

**TABLE 12** Asymptotic efficiency of $\hat{\pi}_{1D}$ relative to $\hat{\pi}_1$ for $m_1 = m_2$ with $\gamma = n/(n+m)$

| | $\Delta$ = 1 | | | $\Delta$ = 2 | | | $\Delta$ = 3 | | |
|---|---|---|---|---|---|---|---|---|---|
| $\gamma$ \ $\pi_1$ | 0.1 | 0.25 | 0.50 | 0.1 | 0.25 | 0.5 | 0.1 | 0.25 | 0.5 |
| 0.1 | 57.2 | 68.3 | 74.0 | 56.9 | 77.5 | 85.8 | 71.0 | 88.6 | 94.3 |
| 0.25 | 58.4 | 70.8 | 77.5 | 54.3 | 77.5 | 87.6 | 68.0 | 88.0 | 94.3 |
| 0.5 | 58.9 | 74.9 | 84.4 | 47.4 | 75.4 | 89.9 | 60.0 | 85.0 | 94.2 |
| 0.75 | 55.5 | 78.5 | 91.5 | 34.9 | 66.0 | 85.1 | 44.0 | 74.3 | 88.7 |
| 0.9 | 45.8 | 77.2 | 88.3 | 21.5 | 45.8 | 61.3 | 24.9 | 51.3 | 69.7 |

Reproduced from S. Ganesalingam and G.J. McLachlan (1981), Some efficiency results for the estimation of the mixing proportion in a mixture of two normal distributions. *Biometrics* **37**, 23-33. With permission from the Biometric Society.

contains a high proportion of unclassified observations—not an uncommon occurrence since there are generally only a limited number of classified observations available. In other situations the asymptotic relative efficiency of $\hat{\pi}_{1D}$ can be quite high provided that $\gamma$ is not too close to one ($\gamma \leq 0.8$, say).

A series of simulation experiments was performed by Ganesalingam and McLachlan (1981) in order to assess to what extent the asymptotic efficiency results for $\hat{\pi}_{1D}$ relative to $\hat{\pi}_1$ are applicable in situations where the sample sizes $m$ and $n$ are not very large. It was found that there was generally reasonable agreement between the asymptotic and simulated efficiencies, although the overall agreement was clearly better for equal than for disparate mixing proportions. Also, for combinations where $p$ was not small relative to $m$, the simulated relative efficiences were higher than the asymptotic predictions, which was subsequently found to be due to the asymptotic variance of $\hat{\pi}_1$ being an underapproximation of the true variance in these circumstances. A few simulations were performed also by Ganesalingam and McLachlan (1981) for normal populations with proportional covariance matrices, $\mathbf{\Sigma}_2 = \kappa \mathbf{\Sigma}_1$ ($\kappa > 1$), to illustrate the asymptotic relative efficiency of $\hat{\pi}_{1D}$ for a mixture of normal heteroscedastic components. The parameter

$$\Delta^* = \Delta / \left\{ \tfrac{1}{2}(1 + \sqrt{\kappa}) \right\}$$

was introduced as a measure of the degree of separation between $G_1$ and $G_2$, where now

$$\Delta^2 = (\boldsymbol{\mu}_1 - \boldsymbol{\mu}_2)' \mathbf{\Sigma}_1^{-1} (\boldsymbol{\mu}_1 - \boldsymbol{\mu}_2).$$

The rule $R$ was based on the usual sample quadratic discriminant function, and its allocation rates were estimated using cross-validation. The sampling distribution of the quadratic discriminant function is too complicated for manageable analytical expressions to exist for its allocation rates, although some progress has been made in providing asymptotic expansions in the case of proportional covariance matrices (see, for example, McLachlan, 1975c). The simulated values of the relative efficiency of $\hat{\pi}_{1D}$ were generally lower than those obtained previously for comparable combinations of the parameters in the homoscedastic case. Some drop in the performance of $\hat{\pi}_{1D}$ was anticipated, however, since in forming $\hat{\pi}_{1D}$, the conditional allocation rates were estimated nonparametrically in contrast to the parametric approach used under homoscedasticity.

## 4.5 MOMENT ESTIMATORS

For a mixture of two populations with means $\boldsymbol{\mu}_1$ and $\boldsymbol{\mu}_2$ and covariance matrices $\boldsymbol{\Sigma}_1$ and $\boldsymbol{\Sigma}_2$ respectively, the mean and covariance matrix of the mixture are given by

$$\boldsymbol{\mu} = \pi_1\boldsymbol{\mu}_1 + \pi_2\boldsymbol{\mu}_2$$

and

$$\mathbf{V} = \pi_1\boldsymbol{\Sigma}_1 + \pi_2\boldsymbol{\Sigma}_2 + \pi_1\pi_2(\boldsymbol{\mu}_1 - \boldsymbol{\mu}_2)(\boldsymbol{\mu}_1 - \boldsymbol{\mu}_2)'.$$

Let $\bar{\mathbf{x}}$ be the sample mean of the $n$ unclassified observations $\mathbf{x}_j$ ($j = 1, \ldots, n$). Walker (1980) proposed

$$\hat{\pi}_{1M} = \{(\boldsymbol{\mu}_1 - \boldsymbol{\mu}_2)'\mathbf{V}^{-1}(\bar{\mathbf{x}} - \boldsymbol{\mu}_2)\} \,/\, \{(\boldsymbol{\mu}_1 - \boldsymbol{\mu}_2)'\mathbf{V}^{-1}(\boldsymbol{\mu}_1 - \boldsymbol{\mu}_2)\}$$

as an estimator of $\pi_1$. He showed that of the unbiased estimators of $\pi_1$ of the form

$$\{M(\bar{\mathbf{x}}) - M(\boldsymbol{\mu}_2)\} \,/\, \{M(\boldsymbol{\mu}_1) - M(\boldsymbol{\mu}_2)\},$$

$\hat{\pi}_{1M}$ has minimum variance among all linear maps $M$ from $\mathbf{R}^p$ to $\mathbf{R}$ such that $M(\boldsymbol{\mu}_1) \neq M(\boldsymbol{\mu}_2)$; $\hat{\pi}_{1M}$ corresponds to $M$ defined by

$$M(\mathbf{x}) = \mathbf{x}'\mathbf{V}^{-1}(\boldsymbol{\mu}_1 - \boldsymbol{\mu}_2). \tag{4.5.1}$$

The estimator $\hat{\pi}_{1M}$ can take on negative values but this can be rectified by appropriate truncation at zero or one. Unfortunately, for $p > 1$, $\hat{\pi}_{1M}$ depends on $\pi_1$ through $\mathbf{V}$. One way of proceeding is to replace $\mathbf{V}$ by $\mathbf{S}_x$, the sample covariance matrix of the unclassified observations $\mathbf{x}_j$ ($j = 1, \ldots, n$). In practice the means of the component distributions are usually unknown but, if there are data $\mathbf{y}_{ij}$ ($j = 1, \ldots, m_i$), of known origin from $G_i$ ($i = 1, 2$), then $\boldsymbol{\mu}_1$ and $\boldsymbol{\mu}_2$ can be estimated by $\bar{\mathbf{y}}_1$ and $\bar{\mathbf{y}}_2$. This leads to the sample version of $\hat{\pi}_{1M}$,

$$\hat{\pi}_{1M} = \{(\bar{\mathbf{y}}_1 - \bar{\mathbf{y}}_2)'\mathbf{S}_x^{-1}(\bar{\mathbf{x}} - \bar{\mathbf{y}}_2)\} \,/\, \{(\bar{\mathbf{y}}_1 - \bar{\mathbf{y}}_2)'\mathbf{S}_x^{-1}(\bar{\mathbf{y}}_1 - \bar{\mathbf{y}}_2)\}, \tag{4.5.2}$$

which can be viewed as the moment estimator of $\pi_1$ after transformation of the original data according to (4.5.1).

Besides being very easy to compute, the estimator $\hat{\pi}_{1M}$ does not require knowledge of the forms of the component distributions. However, for a mixture of univariate normal distributions with known parameters, James (1978) observed from his asymptotic and simulative studies that $\hat{\pi}_{1M}$ is quite inefficient relative to the likelihood estimator $\hat{\pi}_1$ except when the overlap between the distributions is very severe. It will be seen that a similar result holds for the relative efficiency of $\hat{\pi}_{1M}$ for multivariate normal distributions with unknown parameters. Hence, except for the purpose of providing a starting value for $\hat{\pi}_1$ in its iterative computation, the use of $\hat{\pi}_{1M}$ would appear to be more relevant in situations where the forms of the component distributions are unknown.

For a mixture of two multivariate normal homoscedastic populations, McLachlan (1982b) derived the asymptotic bias and variance of $\hat{\pi}_{1M}$ as given by (4.5.2), along with its asymptotic relative efficiency. The latter is displayed as a percentage in Table 13 for various combinations of the parameters $\pi_1$, $\Delta$, and $\gamma = n/(m+n)$ with $m_1 = m_2$; it does not depend explicitly on $p$. Also, since $m_1 = m_2$, interchanging $\pi_1$ with $1 - \pi_1$ does not alter the entries in Table 13, which has been taken from McLachlan (1982b).

It can be seen that the asymptotic relative efficiency of $\hat{\pi}_{1M}$ is not high except for the three combinations with $\pi_1 = 0.5$, $\Delta = 1$, and $\gamma \leq 0.5$, representing distributions with considerable overlap in equal proportions with the combined sample having no more than 50% of the observations unclassified. A comparison of Table 13 with the corresponding values of the asymptotic efficiency of the discriminant analysis estimator $\hat{\pi}_{1D}$ relative to $\hat{\pi}_1$ in Table 12 suggests that $\hat{\pi}_{1M}$ is more efficient than $\hat{\pi}_{1D}$ essentially only for distributions very close together providing $\gamma \leq 0.5$. However, it should

**TABLE 13** Asymptotic efficiency of $\hat{\pi}_{1M}$ relative to $\hat{\pi}_1$ for $m_1 = m_2$ with $\gamma = n/(n+m)$

| | $\Delta$ | | | | | | | | |
|---|---|---|---|---|---|---|---|---|---|
| | 1 | | | 2 | | | 3 | | |
| | $\pi_1$ | | | | | | | | |
| $\gamma$ | 0.1 | 0.25 | 0.50 | 0.1 | 0.25 | 0.5 | 0.1 | 0.25 | 0.5 |
| 0.1 | 76.7 | 90.8 | 97.8 | 59.7 | 80.9 | 89.4 | 57.4 | 77.4 | 84.3 |
| 0.25 | 72.4 | 88.4 | 97.3 | 52.4 | 76.2 | 87.0 | 49.6 | 71.8 | 81.0 |
| 0.5 | 64.7 | 83.9 | 95.8 | 39.3 | 64.0 | 79.3 | 35.2 | 59.4 | 70.9 |
| 0.75 | 54.5 | 77.8 | 91.2 | 24.8 | 47.5 | 62.4 | 19.1 | 38.8 | 50.8 |
| 0.9 | 42.3 | 70.7 | 80.2 | 13.8 | 28.5 | 37.3 | 8.8 | 19.2 | 27.1 |

Reprinted from G.J. McLachlan (1982b), *Commun. Statist.-Simula. Computa.* **11**, 715-726. By courtesy of Marcel Dekker, Inc.

be noted that this version of $\hat{\pi}_{1D}$ is not as robust against normality as $\hat{\pi}_{1M}$ since the discriminant rule and its estimated allocation rates used in the formation of $\hat{\pi}_{1D}$ were appropriate for the underlying model of normal component distributions with equal covariance matrices.

## 4.6 MINIMUM DISTANCE ESTIMATORS

The likelihood, discriminant analysis, and moment estimators of the mixing proportions can all be obtained by using the method of minimum distance through an appropriate choice of the distance measure. For a review of the properties of this method in a general estimation context the reader is referred to Parr (1981), who has compiled an extensive bibliography, and Beran (1984). Under suitable regularity conditions minimum distance estimators will have the desirable asymptotic properties of consistency and normality.

In the present mixture context the vector of unknown parameters, $\phi = (\pi', \theta')'$ is estimated by $\hat{\phi}_{MD}$, obtained by minimizing

$$\delta(\hat{F}, F), \tag{4.6.1}$$

the distance between the mixture distribution function $F$ and the empirical distribution function $\hat{F}$ based on $\mathbf{x}_1, \ldots, \mathbf{x}_n$ drawn from the mixture $G$. Titterington, Smith, and Makov (1985, Chapter 4) have provided a comprehensive account of the properties of minimum distance estimators for mixtures, in particular, for the estimation of the mixing proportions. They described various distance measures $\delta$ which have been used, including those where densities rather than distribution functions are used in (4.6.1). It was noted that this class includes likelihood estimators if $\delta$ is taken to be the Kullback-Leibler directed divergence, where

$$\delta(\hat{F}, F) = \int \log\{d\hat{F}(\mathbf{x})/dF(\mathbf{x})\} d\hat{F}(\mathbf{x}).$$

For univariate data, Choi (1969a) established the strong consistency and asymptotic normality of $\hat{\phi}_{MD}$ under the typical sort of regularity conditions for the quadratic version of (4.6.1) given by

$$\begin{aligned} \delta(\hat{F}, F) &= \int \left\{F(x;\phi) - \hat{F}(x)\right\}^2 d\hat{F}(x) \\ &= \sum_{j=1}^{n} \{F(x_{(j)};\phi) - (j/n)\}^2/n, \end{aligned} \tag{4.6.2}$$

where $x_{(j)}$ denotes the $j$th order statistic ($j = 1, \ldots, n$). A treatment of the practical aspects of this problem may be found in Macdonald and Pitcher (1979). Choi and Bulgren (1968) had used (4.6.2) for the estimation of the mixing proportions where the component distributions were known; see also Choi (1969b). Macdonald (1971) subsequently provided empirical evidence that a less biased estimator of the mixing proportion $\pi_1$ is obtained in a mixture of two normal distributions if $\delta$ is taken to be the Cramér-von Mises distance,

$$\delta(\hat{F}, F) = \int \{F(x;\phi) - \hat{F}(x)\}^2 dF(x;\phi)$$

$$= \tfrac{1}{12}n^{-2} + \sum_{j=1}^{n}\{F(x_{(j)};\phi) - (j - \tfrac{1}{2})/n\}^2/n.$$

Recently, for this distance measure, Woodward, Parr, Schucany, and Lindsey (1984) considered the properties of the estimator $\hat{\phi}_{MD}$ in relation to the likelihood estimator $\hat{\phi}$ formed under the assumption of a mixture of two normal univariate densities with unknown means and variances. Their attention was focused on the estimation of the mixing proportion $\pi_1$. A simulation study revealed that $\hat{\phi}$ is superior to $\hat{\phi}_{MD}$ when normality holds for the component populations, but that $\hat{\phi}_{MD}$ is more robust to symmetric departures from component normality, providing better estimates for mixtures of heavy-tailed densities. Strong consistency and asymptotic normality were established under regularity conditions that are satisfied for normal component populations.

In the case where the mixing proportions are the only unknown parameters or where there are classified data available to provide estimates of the unknown parameters in the component distributions, then $F$ is linear in the $\pi_i$ and so an explicit expression for the estimate of $\boldsymbol{\pi}$ can be obtained if $\delta$ is a linear or quadratic based measure. Concerning quadratic based versions of (4.6.1), it was noted in Section 4.3, on the development of the discriminant analysis estimator of the mixing proportions, that Macdonald (1975) has surveyed the class of weighted least squares estimators for known component distribution functions. For univariate continuous data, Hall (1981) has considered a quadratic based version of (4.6.1) in the case where the component distributions are unknown, but where there are classified data available from each population $G_i$ from which to form the empirical distribution function $\hat{F}_i$ for $i = 1, \ldots, g$. The estimate of $\boldsymbol{\pi}$ was obtained by minimizing

$$\int_{-\infty}^{\infty} \left\{ \hat{F}(x) - \sum_{i=1}^{g} \pi_i \hat{F}_i(x) \right\}^2 u(x)\, dx, \tag{4.6.3}$$

where $u(x)$ is some nonnegative weight function and where (4.6.3) was reparameterized to incorporate the constraint

$$\sum_{i=1}^{g} \pi_i = 1.$$

In the case where the classified data have been sampled from the mixture $G$, Hall (1981) showed that the estimator so obtained is consistent and asymptotically normal. It was indicated that these estimators lead in a natural way to nonparametric forms of well-known parametric estimators. For example, in the case of $g = 2$ for data grouped into two regions $C_1$ and $C_2$ according as $x_j$ is greater or less than some fixed point $k$, (4.6.3) yields

$$\hat{\pi}_{1MD} = \{(\tilde{n}_1/n) - \hat{e}_{21}\} / (\hat{e}_{11} - \hat{e}_{21}), \tag{4.6.4}$$

where now $\hat{e}_{ij}$ is the proportion of the $m_i$ observations in region $C_j$ $(i, j = 1, 2)$. That is, (4.6.4) is the version of the discriminant analysis estimator (4.3.3) where the allocation rates $e_{ij}$ are estimated nonparametrically by the apparent allocation rates for the classified data. As demonstrated by Hall (1981), if a weighted $L_1$-norm were used instead of (4.6.1), then we obtain corresponding to $\hat{\pi}_{1M}$, the moment estimate of $\pi_1$ using weighted means,

$$(\bar{x}_u - \bar{y}_{2u})/(\bar{y}_{1u} - \bar{y}_{2u})$$

where

$$\bar{x}_u = \sum_{j=1}^{n} U(x_j)/n$$

and

$$\bar{y}_{iu} = \sum_{j=1}^{m_i} U(y_{ij})/m_i \qquad (i = 1, 2),$$

and $U$ is an indefinite integral of $u$.

Recently, Titterington (1983) adopted a similar approach to Hall (1981), but where the empirical distribution function for the mixture and for each component distribution were replaced by density estimates. The latter were advocated by Titterington (1983) since the extension to the multivariate case is easier and the treatment of discrete data is more natural, particularly when there is no obvious ordering among the points in the sample space. Practical applications were considered for multinomial data both unsmoothed and smoothed and for continuous data smoothed by the kernel method. As in Hall (1981), Titterington (1983) developed the asymptotic theory for the smoothing of multinomial and density functions for the model where the classified data have been obtained by sampling from the mixture $G$.

More recently, Hall and Titterington (1984) derived a Cramér-Rao lower bound for nonparametric estimators of the mixing proportions by constructing a sequence of multinomial approximations and related likelihood estimators. For mixture sampling of the classified data in the same proportions as the $\pi_i$, it was shown in a nonparametric framework that likelihood estimation of the $\pi_i$ and the category probabilities $e_{ij}$ leads to an explicit solution of the estimate of $\pi_i$,

$$\hat{\pi}_{iHT} = \sum_{j=1}^{N} \frac{m_{ij}(m_{\cdot j} + \tilde{n}_j)}{m_{\cdot j}(m+n)} \qquad (i = 1, \ldots, g-1), \tag{4.6.5}$$

where $\tilde{n}_j$ is the number of the unclassified observations $\mathbf{x}_1, \ldots, \mathbf{x}_n$ falling in category $C_j$ and $m_{ij}$ is the number of the $m_i$ classified observations from $G_i$ in $C_j$, and

$$m_{\cdot j} = \sum_{i=1}^{N} m_{ij}.$$

The approximate likelihood estimator so obtained is more efficient than the estimators obtained in their earlier separate work, and is simpler computationally. However, it should be emphasized that (4.6.5) has been obtained under the assumption that the classified data were obtained by sampling from the mixture $G$.

For completely specified component distributions, the estimator of $\boldsymbol{\pi}$ is given explicitly also if the distance measure (4.6.1) is in terms of a transform of the distribution function such as the weighted squared distance between the theoretical and empirical moment generating functions as proposed by

Quandt and Ramsey (1978); see also Schmidt (1982). Bryant and Paulson (1983), like Kumar, Nicklin and Paulson (1979) and others, advocated the use of the characteristic function rather than the moment generating function in this role. On the basis of a mixture of two univariate normal distributions with known means and variances, they concluded that its use leads to an effective procedure for the estimation of the mixing proportions. For a particular form of the weighting (Heathcote, 1977), they showed that their procedure is equivalent to that of minimization with respect to the mixing proportions of the integrated squared error between a density and its kernel estimate.

## 4.7 CASE STUDY

The case study undertaken by Do and McLachlan (1984) is reconsidered here to illustrate likelihood estimation of mixing proportions in a situation where the superpopulation $G$ is a genuine mixture of a finite number of distinct populations. The data set, which was originally supplied by Messrs. R. G. Cunningham and G. Lenton, may be described as follows. There are $g = 7$ species of Malaysian rats on which $p = 4$ variables relating to skull length, teeth row, palatine foramen, and jaw length were measured on their skulls. The number $m_i$ of classified observations from each species $G_i$ is given in Table 14, along with a description of the seven species of rats. The same four variables were subsequently measured on $n = 1,107$ rat skulls collected from owl pellets. The rats constitute part of an owl's diet, and indigestible material is regurgitated as a pellet. The problem is to determine the diet of the owls in terms of the estimated proportion of each species of rat consumed. The sample means and standard deviations of the four variables for the classified reference data from each species $G_i$ and for the unclassified sample are displayed in Table 15.

In keeping with the previous notation, we let $\mathbf{y}_{ij}$ $(j = 1, \ldots, m_i)$ denote the data of known origin from $G_i$ and $\bar{\mathbf{y}}_i$ and $\mathbf{S}_i$ their sample mean and covariance matrix $(i = 1, \ldots, 7)$; $\mathbf{x}_j$ $(j = 1, \ldots, n)$ denote the unclassified observations. This is an example where the classified data have been obtained by sampling separately from each of the specified populations and so provide no information on the unknown proportions.

Do and McLachlan (1984) concluded that it was reasonable to assume that the populations were multivariate normal but heteroscedastic. The extent of the heteroscedasticity can be seen from the standard deviations of the four variates displayed in Table 15 for each of the seven species. In Section 2.5 various methods of assessing the normality assumption for the populations were discussed, including Hawkins' (1981) method which can be used to test simultaneously for normality and homoscedasticity. However,

**TABLE 14** Number of classified observations for each species

| Species of rat ($G_i$) | $m_i$ |
|---|---|
| *Rattus exulans* | 24 |
| *Rattus argentiventer* | 37 |
| *Rattus tiomanicus* | 39 |
| *Rattus annandalfi* | 24 |
| *Rattus rajah* | 16 |
| *Rattus whiteheadi* | 28 |
| *Rattus surifer* | 9 |

Source: Do and McLachlan (1984)

this data set was obviously heteroscedastic, and the assessment of normality for each population $G_i$ ($i = 1, \ldots, 7$) was based on the Mahalanobis squared distances,

$$D(\mathbf{y}_{ij}, \bar{\mathbf{y}}_{i(ij)}; \mathbf{S}_{i(ij)}) \qquad (j = 1, \ldots, m_i),$$

formed using the individual population sample covariance matrix instead of the pooled version; $\bar{\mathbf{y}}_{i(ij)}$ and $\mathbf{S}_{i(ij)}$ denote the sample mean and covariance

**TABLE 15** Sample means and standard deviations for the classified data by species and the unclassified sample (all measurements are in mm $\times$ $10^{-1}$)

| Source | Means | | | | Standard deviations | | | |
|---|---|---|---|---|---|---|---|---|
| | $p$ | | | | $p$ | | | |
| | 1 | 2 | 3 | 4 | 1 | 2 | 3 | 4 |
| $G_1$ | 311.8 | 47.5 | 52.7 | 117.6 | 12.8 | 1.6 | 3.8 | 5.8 |
| $G_2$ | 382.8 | 67.9 | 70.9 | 167.5 | 35.4 | 2.8 | 9.0 | 19.6 |
| $G_3$ | 375.8 | 61.9 | 65.2 | 156.8 | 19.5 | 2.4 | 5.8 | 13.3 |
| $G_4$ | 418.5 | 75.3 | 70.0 | 182.0 | 32.0 | 1.6 | 6.7 | 17.4 |
| $G_5$ | 386.4 | 65.3 | 56.4 | 147.7 | 25.0 | 2.0 | 5.4 | 13.4 |
| $G_6$ | 323.9 | 51.3 | 39.4 | 121.4 | 9.2 | 1.7 | 3.5 | 5.0 |
| $G_7$ | 422.4 | 61.6 | 64.1 | 158.8 | 26.8 | 2.4 | 4.5 | 12.0 |
| $G$ | 379.8 | 60.5 | 69.1 | 158.7 | 21.3 | 3.1 | 5.0 | 10.4 |

Source: Do and McLachlan (1984)

matrix of the $m_i - 1$ observations $\mathbf{y}_{ik}$ ($k = 1, \ldots, m_i$; $k \neq j$). As seen in Section 2.5,

$$c(m_i, \nu_i) D(\mathbf{y}_{ij}, \bar{\mathbf{y}}_{i(ij)}; \mathbf{S}_{i(ij)}) \tag{4.7.1}$$

can be computed as

$$\frac{(\nu_i m_i / p) D(\mathbf{y}_{ij}, \bar{\mathbf{y}}_i; \mathbf{S}_i)}{(\nu_i + p)(m_i - 1) - m_i D(\mathbf{y}_{ij}, \bar{\mathbf{y}}_i; \mathbf{S}_i)},$$

where $\nu_i = m_i - p - 1$ and $c(.,.)$ is defined by (2.5.6). Under the assumption of normality, the statistic (4.7.1) is distributed according to the $F_{p,\nu_i}$ distribution. Do and McLachlan (1984) reported that $Q$-$Q$ plots constructed for each population on the basis of the $F$ distribution for (4.7.1) revealed no strong evidence against normality. For the seventh population there are only nine classified observations, and so normality cannot be confirmed with any degree of certainty.

Further confirmation of the normality assessment is provided by the nonsignificance of each Anderson-Darling statistic computed for the tail areas $a_{ij}$ ($j = 1, \ldots, m_i$) for each $i$ ($i = 1, \ldots, 7$), where $a_{ij}$ is the area to the right of the observed value of (4.7.1) under the $F_{p,\nu_i}$ distribution. There are some atypical observations among the classified data as indicated by some small values of the $a_{ij}$. To assess the influence of any such outliers, robust estimates of the population means and covariance matrices were computed, using the Huber $M$-estimates of $\boldsymbol{\mu}_i$ and $\boldsymbol{\Sigma}_i$ as given by (2.8.2) and (2.8.7). They were found, however, to be very similar to $\bar{\mathbf{y}}_i$ and $\mathbf{S}_i$ ($i = 1, \ldots, 7$).

The next part of the analysis concerns the treatment of the unclassified observations $\mathbf{x}_j$ ($j = 1, \ldots, n$) in conjunction with the classified data. The maximum likelihood estimates of the $\pi_i$, $\boldsymbol{\mu}_i$, and the $\boldsymbol{\Sigma}_i$ on the basis of the classified and unclassified data combined exist, as there are more than $p = 4$ classified observations from each of the seven specified populations. The application of the EM algorithm to the present problem to compute the maximum likelihood estimates on the basis of all the data would seem to be particularly appropriate, as there is only a limited number of classified observations ($m = 177$), but a very large number of unclassified observations ($n = 1,107$). Moreover, it will be seen that the discriminant analysis approach gives disparate estimates of the mixing proportions. As reported in Section 2.3 on the asymptotic relative efficiency of the discriminant analysis estimator, it is in situations like this there is a good deal to be gained in using the asymptotically efficient likelihood estimator of $\boldsymbol{\pi}$. The maximum likelihood estimates of $\boldsymbol{\pi}$, $\boldsymbol{\mu}_i$, and $\boldsymbol{\Sigma}_i$ $(i = 1, \ldots, 7)$ can be computed iteratively from (2.1.2), (2.1.8), and (2.1.9).

Before proceeding, however, with the computation of these estimates, the typicality of the unclassified data with respect to the seven specified populations $G_1, \ldots, G_7$ is considered. The tail area, $a_{i,j}$, to the right of the observed value of

$$c(m_i + 1, \nu_i + 1)D(\mathbf{x}_j, \bar{\mathbf{y}}_i; \mathbf{S}_i)$$

under the $F_{p,\nu_i+1}$ distribution, provides an assessment of how typical each unclassified observation $\mathbf{x}_j$ is of $G_i$. The observation $\mathbf{x}_j$ can then be assessed as atypical of the mixture $G$ of $G_1, \ldots, G_7$ if $a_j$ is less than some threshold $\alpha$, where

$$a_j = \max_i a_{i,j}.$$

As discussed in Section 2.6, $a_j$ can be regarded as the $P$-value for a test of the compatibilty of $\mathbf{x}_j$ with the mixture $G$. For the present data set there are 153 observations $\mathbf{x}_j$ with $a_j$ less than 0.05 and 255 with $a_j < 0.1$.

In some applications one explanation for the presence of atypical observations is that they may actually come from other populations not specified in the model. However, in this example, the seven specified species are thought to provide a complete representation. It is speculated that the excess of atypical observations among the unclassified data may be due to substantial deformation of some of the rat skulls during their regurgitation in the form of pellets or their subsequent erosion prior to measurement.

In order to eliminate any undue influence that atypical observations may have in the estimation process, Do and McLachlan (1984) first deleted the observations $\mathbf{x}_j$ with $a_j < 0.05$ before undertaking maximum likelihood estimation. They settled on this value of $\alpha$ after noting in particular that the maximum likelihood estimates of the $\boldsymbol{\mu}_i$ and the $\boldsymbol{\Sigma}_i$ on the basis of the classified data and those unclassified observations retained at level $\alpha$ did not change to any marked degree when $\alpha$ was increased to 0.1.

In Table 16, we display the maximum likelihood estimates, $\hat{\pi}_i$, of the mixing proportions $\pi_i$ on the basis of the classified data combined with those 954 unclassified observations $\mathbf{x}_j$ for which $a_j \geq 0.05$. Also displayed in this table are the corresponding estimates of the $\pi_i$ as given by the discriminant analysis method and a robust version of the mixture likelihood approach. The discriminant analysis estimates, $\hat{\pi}_{iD}$, were computed according to (4.3.5) with the discriminant rule $R$ defined by (4.3.7), where the population densities were taken to be normal with mean $\bar{\mathbf{y}}_i$ and covariance matrix $\mathbf{S}_i$ $(i = 1, \ldots, 7)$. The rule $R$ was applied to the same

954 unclassified observations used in conjunction with the classified data in forming the maximum likelihood estimates. The associated allocation rates $e_{ij}$ were estimated by the apparent rates for $R$ trained on the classified data. The estimate of $\pi_4$ so obtained was negative and was taken to be zero. The estimates $\hat{\pi}_i$ and $\hat{\pi}_{iD}$ in Table 16 are slightly different to those in Do and McLachlan (1984), who inadvertently reported the results in the case where 24 unclassified observations additional to the 153 with $a_j < 0.05$ were first deleted.

The estimates of the mixing proportions $\pi_i$ in Table 16 suggest that the rat diet of the owls is composed mainly of the third species with the remainder of their diet consisting mostly of the second species. Contrasting the discriminant analysis estimates with the maximum likelihood solutions for the mixing proportions we see that the effect of maximum likelihood estimation is to increase the estimate of $\pi_3$ by about 3 percent while reducing the estimate of $\pi_2$ by approximately one fifth and the estimate of $\pi_5$ to zero; the estimates of the other proportions are or almost are the same.

The sample mean $\bar{\mathbf{y}}_i$ of the skull characteristics based on the classified data from species $G_i$ ($i = 1, \ldots, 7$) is represented in Figure 6, using the mapping

$$\bar{\mathbf{y}}_i \longrightarrow (\bar{\mathbf{y}}_i)_1/\sqrt{2} + (\bar{\mathbf{y}}_i)_2 \sin t + (\bar{\mathbf{y}}_i)_3 \cos t + (\bar{\mathbf{y}}_i)_4 \sin 2t, \quad -\pi \le t \le \pi,$$

introduced by Andrews (1972). Also displayed in Figure 6 is the corresponding curve for the sample mean of the unclassified observations $\mathbf{x}_j$ with $a_j \ge 0.05$. It can be seen that the curves for this subset of the unclassified data and the third species are very close together over the entire

**TABLE 16** Estimates of mixing proportions

| | Method | |
|---|---|---|
| | Discriminant analysis | Mixture likelihood |
| Species number $i$ | $\hat{\pi}_{iD}$ | $\hat{\pi}_i$ (Robust version) |
| 1 | 0.001 | 0.001 (0.003) |
| 2 | 0.072 | 0.058 (0.063) |
| 3 | 0.897 | 0.925 (0.926) |
| 4 | 0.000 | 0.000 (0.000) |
| 5 | 0.017 | 0.000 (0.000) |
| 6 | 0.000 | 0.000 (0.000) |
| 7 | 0.013 | 0.016 (0.008) |

range $-\pi \leq t \leq \pi$. This is consistent with the implication of the estimates of the mixing proportions that the unclassified set of skulls can be regarded as being almost entirely from the third species.

In Section 2.8 we described a robust estimation procedure with mixture models, whereby observations assessed as atypical of a component population or of the mixture itself are automatically given reduced weight in the computation of the estimates of the unknown parameters. We used Huber's $\psi$-function here to compute robust estimates of the $\boldsymbol{\mu}_i$, $\Sigma_i$, and the $\pi_i$ from (2.8.13), (2.8.14), and (2.8.16). The tuning constant $k_1(p)$ in the definition (2.8.5) of the $\psi$-function was given by (2.8.9) with $p = 4$. Given the excess of highly atypical observations among the unclassified data, it was decided to delete beforehand all unclassified observations $\mathbf{x}_j$ for which $a_j < 0.01$. The estimates of the mixing proportions obtained after so fitting a mixture model robustly to the remaining unclassified 1,090 observations in conjunction with the classified data are listed in parentheses in Table 16. It can be seen that the effect of retaining those unclassified observations $\mathbf{x}_j$ with $0.01 \leq a_j < 0.05$, which were previously deleted, and of giving reduced weight to them as well as to any atypical observations among the classified data, is to increase slightly the estimates of $\pi_1$ to $\pi_3$ with a corresponding decrease in the estimate of $\pi_7$.

## 4.8 HOMOGENEITY OF MIXING PROPORTIONS

As explained by Choi (1979), mixture models are of relevance in a life testing situation, where there are two causes for failure which act in a mutually exclusive manner. The density of the random variable representing failure or survival time can be modeled as

$$f(x;\phi) = \pi_1 f_1(x;\theta) + \pi_2 f_2(x;\theta), \tag{4.8.1}$$

where $f_i(x;\theta)$ is the density for failure time solely due to cause $i$ ($i = 1,2$). Choi (1979) used the mixture model (4.8.1) to compare the toxicity of two chemical agents used in chemotherapy. It was assumed that death was attributable either to toxicity of the agent or to regrowth of the tumour; the toxic death usually precedes the passage due to regrowth.

Accordingly, the mixture $G^{(k)}$ with density

$$f(x;\phi_k) = \pi_{1k} f_1(x;\theta) + \pi_{2k} f_2(x;\theta),$$

was used to represent time to death under chemical agent $k$ ($k = 1,2$), where $\phi_k = (\boldsymbol{\pi}_k', \theta')'$ and $\boldsymbol{\pi}_k = (\pi_{1k}, \pi_{2k})'$. This led to the problem of

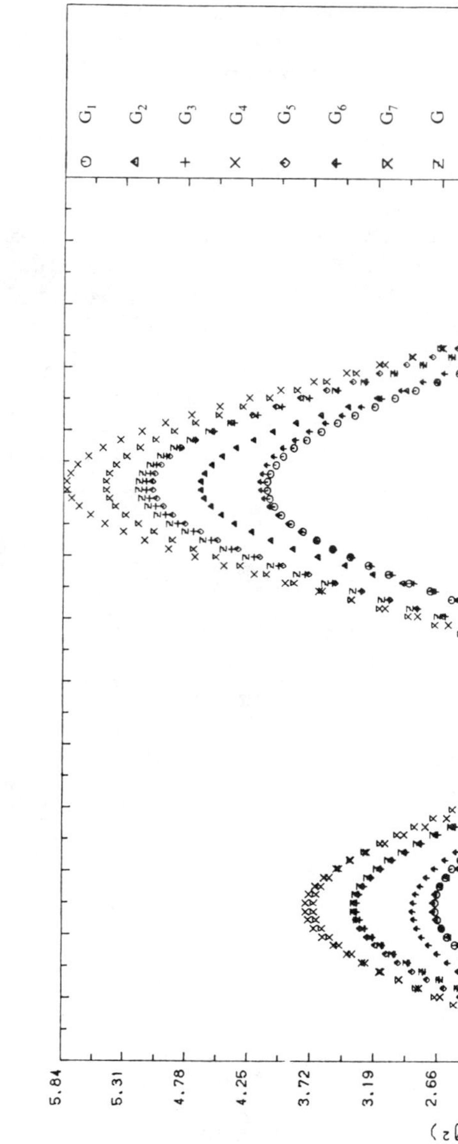

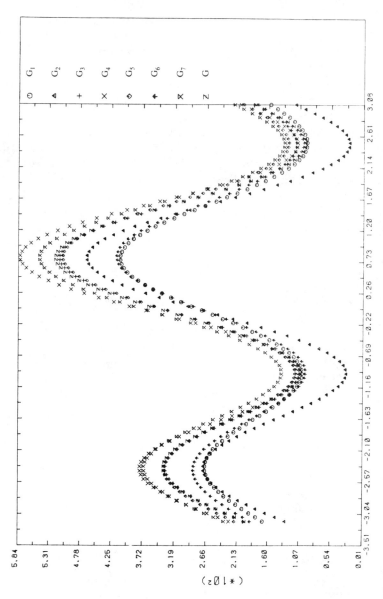

**FIGURE 6** Andrew's sine curves for the mean skull characteristics of each of seven species and of the unclassified sample

testing whether the two mixtures were identical; that is, whether

$$H_0 : \pi_1 = \pi_2 \tag{4.8.2}$$

holds. There were data $x_{jk}$ $(j = 1, \ldots, n_k)$ available from the mixture $G^{(k)}$ $(k = 1,2)$. Note that elsewhere in this book $n_k$ will always denote the number of observations from the $k$th component $G_k$ of a mixture. For univariate normal and exponential densities, Choi (1979) presented four methods for testing $H_0$. Two were parametric tests based on the asymptotic normality of two unbiased estimators of $\pi_{11} - \pi_{12}$, while the other two were nonparametric based on the $U$ and Kolmogorov-Smirnov statistics respectively.

We describe here the likelihood ratio test put forward by McLachlan, Lawoko and Ganesalingam (1982) for testing $H_0$. This test should have more power, at least asymptotically, and also it can handle multivariate data. Since the null hypothesis (4.8.2) is specified in the interior of the parameter space, regularity conditions for the asymptotic null distribution of the likelihood ratio statistic do not break down as with null hypotheses concerning the number of distinct components in the mixture.

Our formulation here is for mixtures of two multivariate normal densities with different means and covariance matrices. Let $\boldsymbol{\phi}^+ = (\boldsymbol{\pi}_1', \boldsymbol{\pi}_2', \boldsymbol{\theta}')'$, and $\boldsymbol{\phi}_k = (\boldsymbol{\pi}_k', \boldsymbol{\theta}')'$ for $k = 1,2$; $\boldsymbol{\theta}$ contains the elements of $\boldsymbol{\mu}_i$ and the distinct elements of $\boldsymbol{\Sigma}_i$ $(i = 1,2)$. In accordance with our previous notation, let $\boldsymbol{\phi} = (\boldsymbol{\pi}', \boldsymbol{\theta}')'$, where now $\boldsymbol{\pi}$ is the common value of $\boldsymbol{\pi}_1$ and of $\boldsymbol{\pi}_2$ under $H_0$. The log likelihood function for $\boldsymbol{\phi}^+$ is given by

$$L_1(\boldsymbol{\phi}_1) + L_2(\boldsymbol{\phi}_2),$$

where $L_k(\boldsymbol{\phi}_k)$ is the log likelihood for $\boldsymbol{\phi}_k$ formed from the $n_k$ observations $\mathbf{x}_{jk}$ $(j = 1, \ldots, n_k)$ from mixture $G^{(k)}$ $(k = 1,2)$. The likelihood ratio test rejects $H_0$ at nominal level $\alpha$ if

$$2\{L_1(\tilde{\boldsymbol{\phi}}_1) + L_2(\tilde{\boldsymbol{\phi}}_2) - L_1(\hat{\boldsymbol{\phi}}) - L_2(\hat{\boldsymbol{\phi}})\} > c, \tag{4.8.3}$$

where $c$ denotes the $(1 - \alpha)$th quantile of the chi-squared distribution with one degree of freedom. In (4.8.3), $\hat{\boldsymbol{\phi}}$ denotes the likelihood estimate of $\boldsymbol{\phi}$ under $H_0$ and $\tilde{\boldsymbol{\phi}}_k = (\tilde{\boldsymbol{\pi}}_k', \tilde{\boldsymbol{\theta}}')'$ for $k = 1,2$, where $\tilde{\boldsymbol{\phi}}^+ = (\tilde{\boldsymbol{\pi}}_1', \tilde{\boldsymbol{\pi}}_2', \tilde{\boldsymbol{\theta}}')'$ is the likelihood estimate of $\boldsymbol{\phi}^+$ in the unconstrained case. Let

$$\hat{\tau}_{ijk} = \tau_i(\mathbf{x}_{jk}; \hat{\boldsymbol{\phi}})$$

and

$$\tilde{\tau}_{ijk} = \tau_i(\mathbf{x}_{jk}; \tilde{\phi}_k)$$

be the estimated posterior probability that $\mathbf{x}_{jk}$ belongs to the $i$th component $G_i$ $(i = 1{,}2)$ under $H_0$ and in the unconstrained case, respectively $(k = 1, 2;\ j = 1, \ldots, n_k)$.

It follows on application of the EM algorithm that the likelihood estimate $\tilde{\phi}^+$ can be obtained iteratively from

$$\tilde{\pi}_{ik} = \sum_{j=1}^{n_k} \tilde{\tau}_{ijk}/n_k \qquad (k = 1, 2),$$

$$\tilde{\boldsymbol{\mu}}_i = \sum_{k=1}^{2}\sum_{j=1}^{n_k} \tilde{\tau}_{ijk}\mathbf{x}_{jk}/(n_1\tilde{\pi}_{i1} + n_2\tilde{\pi}_{i2}), \tag{4.8.4}$$

and

$$\tilde{\boldsymbol{\Sigma}}_i = \sum_{k=1}^{2}\sum_{j=1}^{n_k} \tilde{\tau}_{ijk}(\mathbf{x}_{jk} - \tilde{\boldsymbol{\mu}}_i)(\mathbf{x}_{jk} - \tilde{\boldsymbol{\mu}}_i)'/(n_1\pi_{i1} + n_2\pi_{i2}) \tag{4.8.5}$$

for $i = 1{,}2$. On the computation of $\hat{\phi}$ under $H_0$, we have that

$$\hat{\boldsymbol{\pi}} = (n_1\tilde{\boldsymbol{\pi}}_1 + n_2\tilde{\boldsymbol{\pi}}_2)/(n_1 + n_2)$$

and $\hat{\boldsymbol{\mu}}_i$ and $\hat{\boldsymbol{\Sigma}}_i$ satisfy (4.8.4) and (4.8.5) with $\tilde{\tau}_{ijk}$ replaced by $\hat{\tau}_{ijk}$.

To illustrate the use of the likelihood ratio statistic for testing the hypothesis $H_0$ of same mixing proportions with common components, McLachlan, Lawoko and Ganesalingam (1982) applied it to some data supplied by Professor S. C. Choi, which is reproduced here in Table 17. The data represent survival times in weeks for two sets of rats which were given dosages of cytoxan at a concentration of 60 mg/kg. The second set was given the full dosage once weekly, while the first set received half the dosage twice weekly. As explained in the introduction of this section, the decision problem (4.8.2) is pertinent to comparing the toxicity of the chemical agent at the two dosage levels.

The likelihood ratio test was applied with the common component distributions of the two mixtures taken to be normal with means $\mu_1$ and $\mu_2$ and variances $\sigma_1^2$ and $\sigma_2^2$. In the unconstrained case, initial values of 0.37, 0.80, 7.0, 13.8, 1.96, and 6.76 for $\pi_{11}$, $\pi_{12}$, $\mu_1$, $\mu_2$, $\sigma_1^2$, and $\sigma_2^2$ led, after 29 iterations, to 0.28, 0.798, 6.81, 13.16, 1.654, and 8.608, respectively, as the

**TABLE 17** Survival time in weeks for two sets of rats

| Set 1 ($n_1 = 40$) | | Set 2 ($n_2 = 44$) | |
|---|---|---|---|
| $x_{j1}$ ($j = 1, 20$) | $x_{j1}$ ($j = 21, 40$) | $x_{j2}$ ($j = 1, 22$) | $x_{j2}$ ($j = 23, 44$) |
| 13.50 | 8.50 | 7.25 | 4.00 |
| 13.25 | 15.00 | 5.25 | 7.50 |
| 8.00 | 8.50 | 6.75 | 9.00 |
| 4.75 | 17.50 | 5.50 | 7.00 |
| 11.50 | 8.75 | 7.25 | 4.75 |
| 5.50 | 16.00 | 6.75 | 7.00 |
| 17.75 | 17.00 | 7.50 | 8.75 |
| 8.50 | 15.75 | 6.50 | 7.50 |
| 10.50 | 15.00 | 7.50 | 7.50 |
| 12.75 | 11.50 | 7.00 | 7.75 |
| 12.00 | 13.50 | 8.75 | 7.50 |
| 12.75 | 6.75 | 12.25 | 7.25 |
| 9.25 | 6.75 | 12.50 | 16.00 |
| 5.50 | 6.75 | 6.50 | 14.50 |
| 5.25 | 5.00 | 7.00 | 6.75 |
| 11.50 | 18.00 | 6.50 | 16.00 |
| 11.25 | 8.50 | 7.00 | 12.50 |
| 12.50 | 13.00 | 10.00 | 6.75 |
| 11.75 | 17.75 | 5.50 | 18.00 |
| 13.50 | 11.00 | 7.00 | 5.00 |
| | | 10.00 | 7.50 |
| | | 7.25 | 6.50 |

Source: McLachlan, Lawoko and Ganesalingam (1982).

unconstrained estimates. Under $H_0$, starting with the same initial values (apart from a common value of $\pi_1 = 0.595$ for $\pi_{11}$ and $\pi_{12}$) led, after 22 iterations, to 0.564, 6.84, 13.32, 1.662, and 7.964 as the constrained estimates of $\pi_1$, $\mu_1$, $\mu_2$, $\sigma_1^2$, and $\sigma_2^2$ respectively. The iterations were terminated when the change in successive estimates of each parameter was less than 0.001. Various initial values other than those previously mentioned were used to start the iterations, but they all led to the same estimates, and hence value, of the likelihood ratio test statistic $\lambda$. The value of $-2 \log \lambda$ so obtained was equal to 19.2, and so the likelihood ratio test clearly rejected the hypothesis $H_0$ of same mixing proportions in the two mixture distributions for survival time corresponding to the different dosage levels of the chemical agent; from the $\chi_1^2$ distribution, the associated $P$-value is zero for all practical purposes. For this example, the different tests suggested by Choi (1979) also overwhelmingly led to the rejection of $H_0$.

# 5

# Assessing the Performance of the Mixture Likelihood Approach to Clustering

## 5.1 INTRODUCTION

With any application of a cluster analysis technique in practice, there is the problem of assessing the effectiveness of the clustering obtained for the given data set. With the mixture likelihood approach, a measure of the strength of the clustering produced can be formulated in terms of the relative sizes of the estimated posterior probabilities of membership of the components of the mixture. It has been seen with this approach to clustering that the unclassified observations $\mathbf{x}_1, \ldots, \mathbf{x}_n$ are assumed to come from a mixture $G$ of a specified number $g$ of populations $G_1, \ldots, G_g$, and that they can be clustered on the basis of their estimated posterior probabilities $\hat{\tau}_{i1}, \ldots, \hat{\tau}_{in}$ with respect to $G_i$ ($i = 1, \ldots, g$). The observation $\mathbf{x}_j$ is assigned to $G_t$ if

$$\hat{\tau}_{tj} > \hat{\tau}_{ij} \qquad (i = 1, \ldots, g; i \neq t). \tag{5.1.1}$$

If the maximum of $\hat{\tau}_{ij}$ over $i = 1, \ldots, g$ is near to 1 for most of the observations $\mathbf{x}_j$, then it suggests that the mixture approach can cluster the sample at hand into $g$ distinct groups with a high degree of certainty. Conversely, if this maximum is generally well below 1, it indicates that the components of the fitted mixture model are too close together for the sample to be clustered with any certainty. It would be informative, therefore, to have available a quick summary statistic for assessing the overall performance of the mixture approach in clustering the sample at hand. To this end Ganesalingam and McLachlan (1980b) and, more recently, Basford and

McLachlan (1985a) investigated the use of the estimated posterior probabilities in forming estimates of the overall and individual allocation rates of the clustering with respect to the component populations $G_1, \ldots, G_g$ of the mixture model fitted to the data.

In many instances of course the data to be clustered are not from a genuine mixture of some populations. Rather the fitting of a mixture model with $g$ components is just a way of grouping the data into $g$ clusters. In which case the clusters so obtained do not correspond to *a priori* defined populations, but the estimated allocation rates can still be thought of as a measure of the strength of the clustering. They can be interpreted as estimates of the correct allocation rates which would exist if the clustering were assumed, for this particular purpose, to reflect an externally existing partition of the data.

As noted in Section 1.5 on the identifiability of mixture models, there are $g!$ permutations of the component labels for a mixture of $g$ component distributions belonging to the same parametric family. Therefore in such situations when speaking about the correct allocation rates of the clustering produced by the mixture likelihood approach, there is the matter of identifying the $g$ clusters produced with the $g$ external populations. This is not an issue in practice as the estimates of the allocation rates, which are formed under the assumption that the clusters correspond to the appropriate populations, are interpreted solely as a measure of the strength of the clustering. Of course in simulation studies, the clusters need to be identified with the populations $G_1, \ldots, G_g$ if results are to be reported, say, on the agreement between the estimated and actual allocation rates. In the simulation studies of Basford and McLachlan (1985a) to be considered later in this chapter, a cluster was associated with population $G_i$ if the proportion of elements from $G_i$ allocated to this cluster was greater than that for any other cluster. This convention was adopted also in the simulation experiments of Bayne, Beauchamp, Begovich and Kane (1980). There are other ways, as for example in Kuiper and Fisher (1975), who compared various clustering techniques in their application to data generated from two bivariate distributions with means $(0,0)$ and $(\Delta,0)$ for $\Delta > 0$. They identified the cluster in which the sample mean of the first coordinate was the larger with the second population.

In the following sections of this chapter we shall discuss for the clusters produced by the mixture likelihood approach the formation of suitable estimates of the associated allocation rates in terms of the estimated posterior probabilities of population membership. Particular attention is to be given to the role of the bootstrap method of Efron (1979) for correcting the estimated rates for bias. An account is to be presented of the work of Basford and McLachlan (1985a) who investigated their proposal for assessing the performance of the mixture likelihood approach to clustering on the basis of both real and simulated data sets.

## 5.2 ESTIMATORS OF THE ALLOCATION RATES

As explained in the introduction to this chapter, the estimation of the allocation rates is undertaken under the assumption that each unclassified observation $\mathbf{x}_j$ comes from one of the $g$ component populations $G_1, \ldots, G_g$ specified under the mixture model. In this framework, reference shall be made to the true grouping of the unclassified data with respect to these populations. As before, we let $\mathbf{z}_j = (z_{1j}, \ldots, z_{gj})'$ denote the vector of unknown indicator variables associated with a given $\mathbf{x}_j$, where $z_{ij} = 1$, if $\mathbf{x}_j$ belongs to $G_i$, and 0 otherwise. We let $\hat{\mathbf{z}}_j$ $(j = 1, \ldots, n)$ denote the clustering of the $\mathbf{x}_j$ $(j = 1, \ldots, n)$ as obtained on the basis of (5.1.1). That is, if (5.1.1) holds for a given $\mathbf{x}_j$, then $\hat{z}_{ij} = 1$ $(i = t)$ and $\hat{z}_{ij} = 0$ $(i \neq t)$.

The correct allocation rate with respect to $G_i$ is given by

$$A_i = \sum_{j=1}^{n} z_{ij}\delta(z_{ij}, \hat{z}_{ij})/n_i$$

where, for any $u$ and $v$, $\delta(u, v) = 1$ for $u = v$ and 0 for $u \neq v$, and

$$n_i = \sum_{j=1}^{n} z_{ij}$$

denotes the number of observations in the unclassified data $\mathbf{x}_j$ $(j = 1, \ldots, n)$ coming from $G_i$ $(i = 1, \ldots, g)$. The overall correct allocation rate for the mixture $G$ is given by

$$A = \sum_{i=1}^{g} n_i A_i / n.$$

The allocation rates $A_i$ and $A$ depend on the unknown indicator variables $\mathbf{z}_j$ and so must be estimated. This contrasts with the situation in discriminant analysis when the origin of each observation is known, so that these proportions are observable and are referred to as the apparent allocation rates. In the discriminant analysis context, these rates are used to estimate the performance of the allocation rule when applied to subsequent observations of unknown origin.

In the cluster analysis setting, Ganesalingam and McLachlan (1980b) estimated the overall correct allocation $A$ by

$$T = \sum_{j=1}^{n} \max_r \hat{\tau}_{rj}/n. \tag{5.2.1}$$

The reason for their using $T$ as an estimator of $A$ followed from the ideas developed in a discriminant analysis context, by Fukunaga and Kessel (1972, 1973), Lissack and Fu (1976), Glick (1978), Moore, Whitsitt and Landgrebe (1976), and Schwemer and Dunn (1980). In the case of known $\phi$,

$$\sum_{j=1}^{n} \max_r \tau_{rj/n}$$

is a consistent and unbiased estimator of the overall correct rate for the allocation rule based on (5.1.1) using the known posterior probabilities $\tau_{ij}$, and $A$ converges in probability to this latter rate as $n \to \infty$.

Basford and McLachlan (1985a) proposed that the individual correct allocation rates $A_i$ be estimated by

$$T_i = \sum_{j=1}^{n} (\hat{z}_{ij}\hat{\tau}_{ij})/(n\hat{\pi}_i) \qquad (i = 1, \ldots, g). \tag{5.2.2}$$

It can be verified without difficulty that providing $\hat{\phi}$ is consistent under the mixture model, $T - A$ and $T_i - A_i$ ($i = 1, \ldots, g$) converge in probability to zero as $n \to \infty$.

Available results from Ganesalingam and McLachlan (1980b) and Basford and McLachlan (1985a) suggest that the $T_i$ and $T$ tend to overestimate the $A_i$ and $A$, and so some method of bias correction is recommended, even though it may involve considerable computation. Moreover, as explained by Efron (1982), even if correction for bias is not undertaken, it can still be of interest to compute the estimated bias of an estimator, along with an estimate of its standard deviation or root mean squared error (RMSE). If the estimated bias is less than, say $\frac{1}{4}$, of the estimated standard deviation or root mean squared error, then bias is probably not a serious issue. Note, however, that as the quantities being estimated here are not parameters but random variables, the mean squared error does not equate simply to the variance plus the bias squared. For example, considering $T$ as an estimator of $A$, we have that

$$\text{MSE}(T) = \text{var}(T) + \text{var}(A) + \beta^2 - 2\,\text{cov}(T, A),$$

where $\beta = E(T - A)$ is the bias of $T$.

Basford and McLachlan (1985a) considered the use of the $T_i$ and $T$ in assessing the performance of the mixture likelihood approach in its application to each of three real data sets. Each set was obtained by mixing data taken from known sources and so the actual allocation rates of the clusterings produced were known. It was found that $T$ and the $T_i$ generally provided useful information on the clusterings of the sets, although they did give an optimistic assessment, in particular of the overall correct allocation rate. Indeed, the ratio of the estimated bias to the estimated root mean squared error generally well exceeded the suggested threshold of $\frac{1}{4}$. The assessment for one of these sets, the hemophilia data described in Section 3.2, will be discussed in more detail in Section 5.4.

## 5.3 BIAS CORRECTION OF THE ESTIMATED ALLOCATION RATES

Basford and McLachlan (1983, 1985a) investigated the role of the bootstrap method of Efron (1979) in correcting the estimated allocation rates for bias. This method has been used in the previous sections to assess the null distribution of the likelihood ratio test statistic for the number of components in a mixture. It may be applied as follows to the present problem.

Step 1. A new set of data, $\mathbf{X}^* = (\mathbf{x}_1^{*\prime}, \ldots, \mathbf{x}_n^{*\prime})'$ and $\mathbf{Z}^* = (\mathbf{z}_1^{*\prime}, \ldots, \mathbf{z}_n^{*\prime})'$ is generated from the distribution of $(\mathbf{z}', \mathbf{x}')'$ with the estimates $\hat{\boldsymbol{\pi}}$ and $\hat{\boldsymbol{\theta}}$ used in place of the unknown $\boldsymbol{\pi}$ and $\boldsymbol{\theta}$. This can be achieved by generating for each $j$ ($j = 1, \ldots, n$), a random variable which takes the values $1, \ldots, g$ with probabilities $\hat{\pi}_1, \ldots, \hat{\pi}_g$. If the generated value is equal to, say $t$, then $z_{ij}^*$ is set equal to 1 for $i = t$ and to zero otherwise. The observation $\mathbf{x}_j^*$ is then generated from the density $f_t(\mathbf{x}; \hat{\boldsymbol{\theta}})$. This implies that $\mathbf{x}_1^*, \ldots, \mathbf{x}_n^*$ are the observed values of a random sample from the mixture density

$$f(\mathbf{x}; \hat{\boldsymbol{\phi}}) = \sum_{i=1}^{g} \hat{\pi}_i f_i(\mathbf{x}; \hat{\boldsymbol{\theta}}), \tag{5.3.1}$$

and

$$\mathbf{z}_1^*, \ldots, \mathbf{z}_n^* \overset{iid}{\sim} \text{Mult}_g(1, \hat{\pi}), \tag{5.3.2}$$

a multinomial distribution consisting of one draw on $g$ categories with probabilities $\hat{\pi}_1, \ldots, \hat{\pi}_g$ respectively. Note that unlike the indicator vectors $\mathbf{z}_j$ associated with the original observations $\mathbf{x}_j$, the bootstrap indicators $\mathbf{z}_j^*$ are known.

Step 2. The same mixture model as used to cluster the original data $\mathbf{x}_1, \ldots, \mathbf{x}_n$ is now fitted to the bootstrap sample $\mathbf{x}_1^*, \ldots, \mathbf{x}_n^*$ to produce a new estimate of $\phi = (\pi', \theta')'$, say $\hat{\phi}^* = (\hat{\pi}^{*\prime}, \hat{\theta}^{*\prime})'$; knowledge of the $\mathbf{z}_j^*$ is not used in this step.

Step 3. The bootstrap sample is clustered by assigning each $\mathbf{x}_j^*$ on the basis of the maximum value of its estimated posterior probability of (bootstrap) population membership $\tau_i(\mathbf{x}_j^*; \hat{\phi}^*)$ over $i = 1, \ldots, g$. The estimated overall and individual allocation rates $T^*$ and $T_i^*$ $(i = 1, \ldots, g)$ for this clustering are computed, along with the actual (bootstrap) rates $A^*$ and $A_i^*$ $(i = 1, \ldots, g)$, as determined from the known $\mathbf{z}_j^*$.

Step 4. The expectation of the difference $b_i^* = T_i^* - A_i^*$ with respect to the bootstrap distribution given by (5.3.1) and (5.3.2) is referred to as the bootstrap bias and is denoted by $E_*(b_i^*)$ for $i = 1, \ldots, g$. It can be approximated by $\bar{b}_i^*$ obtained by averaging $b_{ik}^*$ over $K$ independently repeated realizations of $\mathbf{X}^*$ and $\mathbf{Z}^*$; that is

$$\bar{b}_i^* = \sum_{k=1}^{K} b_{ik}^*/K, \tag{5.3.3}$$

where $b_{ik}^* = T_{ik}^* - A_{ik}^*$ denotes the value of $T_i^* - A_i^*$ obtained on the $k$th bootstrap replication. The standard error of the Monte Carlo approximation $\bar{b}_i^*$ to the bootstrap bias $E_*(b_i^*)$ is calculated as the positive square root of

$$\sum_{k=1}^{K} (b_{ik}^* - \bar{b}_i^*)^2/K(K-1).$$

The root mean squared error of $T_i$ is estimated by the positive square root of

$$\sum_{k=1}^{K} b_{ik}^{*2}/K.$$

Similarly, the bootstrap estimates of the bias and root mean squared error of $T$ can be formed with respect to its estimation of the overall allocation rate $A$. We let $E_*(b^*)$ denote the bootstrap bias, where $b^* = T^* - A^*$, and $\bar{b}^*$ its Monte Carlo approximation. The bias corrected estimates of the overall and individual allocation rates are given by

$$T^{(B)} = T - \bar{b}^*$$

and

$$T_i^{(B)} = T_i - \bar{b}_i^* \qquad (i = 1, \ldots, g)$$

respectively.

The above assessment of the distribution of the estimated allocation rates provided by the bootstrap can be used to form approximate confidence intervals for the unobservable allocation rates. Efron (1982) has made various suggestions about how the bootstrap method can be used formally to give nonparametric confidence intervals. However, the usefulness of these ideas in the present context has yet to be investigated.

Corresponding to the observation of Efron and Gong (1983) in a discriminant analysis context, the sample variance of the bootstrap replications $b_k^*$ of $b^* = T^* - A^*$,

$$\sum_{k=1}^{K} (b_k^* - \bar{b}^*)^2/(K-1),$$

should provide an estimate of a lower bound on the mean squared error of the bias corrected rate $T^{(B)}$ in its estimation of $A$. This is because the variance of $T - A$ can be viewed as the mean squared error of the "ideal constant" estimator,

$$T^{(\mathrm{IC})} = T - \beta,$$

which would be used if we knew $\beta$, the bias of $T$. To see this note that

$$\begin{aligned}\text{var}(T - A) &= E(T - A - \beta)^2 \\ &= E\{(T - \beta) - A\}^2 \\ &= \text{MSE}(T^{(\text{IC})}).\end{aligned}$$

It would be anticipated that $T^{(B)}$ would have mean squared error at least as large as $T^{(\text{IC})}$. Analogous comments apply with respect to the estimation of the individual correct allocation rates.

The bootstrap procedure as outlined above is the parametric version as used by Basford and McLachlan (1985a). They considered also a semiparametric version using a nonparametric method of generating the bootstrap sample in Step 1, but with Steps 2 to 4 the same as previously outlined. In Step 1, a sample of size $n$ would be drawn with replacement from the observations $\mathbf{x}_1, \ldots, \mathbf{x}_n$ with each observation given equal weight $1/n$. The origin of each chosen observation $\mathbf{x}_j$ would be determined then in accordance with its estimated posterior probability, $\hat{\tau}_{ij}$, of belonging to $G_i$ ($i = 1, \ldots, g$). That is, a random variable which takes the values $1, \ldots, g$ with probabilities $\hat{\tau}_{1j}, \ldots, \hat{\tau}_{gj}$ respectively would be generated; the value of this generated random variable determines the (bootstrap) population of origin of the observation $\mathbf{x}_j$.

In the empirical investigation of Basford and McLachlan (1985a), the number of bootstrap replications was limited to 50 for economical considerations. In various other applications of the bootstrap, Efron (1979, 1981a, 1981b) noted that the choice of replication number usually does not seem to be critical past 50 or 100; see also Efron and Tibshirani (1986), Gong (1986), and Tibshirani (1985). Of course as $K \to \infty$, the standard error of $\bar{b}^*$ and of each $\bar{b}_i^*$ tend to zero, but there is no point in taking $K$ to be any larger than necessary to ensure that the standard error of $\bar{b}^*$ and of each $\bar{b}_i^*$ are small relative to the standard deviation of $E_*(b^*)$ and of each $E_*(b_i^*)$. Some indication about how large $K$ should be to achieve this was given by Basford and McLachlan (1985a) who carried out a components-of-variance analysis for data simulated in four cases, $A$, $B$, $C$, and $D$, as described in Table 18. It showed that as $K \to \infty$ from $K = 50$, the trial-to-trial standard deviation of $\bar{b}_1^*$ would decrease from 0.034, 0.024, 0.014, and 0.020 to 0.029, 0.018, 0.009, and 0.018 in cases $A$, $B$, $C$, and $D$, respectively; similarly, for $\bar{b}_2^*$ and $\bar{b}^*$. Hence in these four cases, $K = 50$ would appear to be adequate, because there would be only a moderate reduction in the standard deviation of $\bar{b}_i^*$ or $\bar{b}^*$ for a larger value of $K$.

The bootstrap procedure described above can be modified without too much effort to give improved estimates of the Monte Carlo approximations.

**TABLE 18** Design aspects of simulation study of Basford and McLachlan (1985a), involving $n$ observations from a mixture in equal proportions of two bivariate normal populations with a common covariance matrix and with Mahalanobis distance $\Delta$ between them

| Case | $\Delta$ | $n$ | Number of trials |
|---|---|---|---|
| A | $\sqrt{2}$ | 60 | 40 |
| B | 2 | 60 | 40 |
| C | 2 | 120 | 20 |
| D | 3 | 60 | 40 |

Step 1, involving the generation of the bootstrap data $\mathbf{X}^*$ and $\mathbf{Z}^*$, is equivalent to the following procedure. For $j = 1, \ldots, n$,

(i) generate $\mathbf{x}_1^*, \ldots, \mathbf{x}_n^*$ according either to the density $f(\mathbf{x}; \hat{\phi})$ in the parametric version or to the probability function with mass $1/n$ at $\mathbf{x}_j$ $(j = 1, \ldots, n)$ in the semiparametric version;
(ii) generate $\mathbf{z}_1^*, \ldots, \mathbf{z}_n^*$ according to

$$\text{Mult}_g(1, \hat{\boldsymbol{\tau}}_j),$$

where $\hat{\boldsymbol{\tau}}_j = (\hat{\tau}_{1j}, \ldots, \hat{\tau}_{gj})'$ is the vector of the estimates of the posterior probabilities for $\mathbf{x}_j$ formed from the original data, that is, $\hat{\tau}_{ij} = \tau_i(\mathbf{x}_j; \hat{\phi})$.

The modification of the bootstrap procedure is achieved by recognizing that the fitting of the mixture model depends on the outcome of (i) only. Therefore, once the mixture model has been fitted for a realization of (i), (ii) and Step 3 can be repeated $M$ times with little additional computational effort to produce new correct allocation rates with respect to the bootstrap populations. That is, the mixture method of clustering is applied only once and the sample is simply "relabeled" a number of times ($M$). Consequently, the estimates of the allocation rates remain the same, but the true rates with respect to the bootstrap populations vary with the relabeling. For example, the Monte Carlo approximation to the bootstrap bias $E_*(b_i^*)$ is given now by

$$\bar{b}_i^* = \sum_{k=1}^{K} \sum_{m=1}^{M} b_{ikm}^*/KM$$

$$= \sum_{k=1}^{K} \bar{b}_{ik}^*/K,$$

where

$$b_{ikm}^* = T_{ik}^* - A_{ikm}^*$$

is the value of $T_i^* - A_i^*$ obtained on the $(k,m)$th bootstrap replication, and

$$\bar{b}_{ik}^* = \sum_{m=1}^{M} b_{ikm}^*/M.$$

For fixed $k$ the $b_{ikm}^*$ are not independent, and so care must be taken when calculating the estimate of the standard error of $\bar{b}_i^*$. It is given by the positive square root of

$$\sum_{k=1}^{K} (\bar{b}_{ik}^* - \bar{b}_i^*)^2/K(K-1).$$

The variance of the Monte Carlo approximation to $E_*(b_i^*)$ is now better with $M > 1$ as the $\bar{b}_{ik}^*$ have smaller variance than the $b_{ik}^*$ used in the bootstrap algorithm initially described in this section. Note, however, that it is not possible to get an estimate with zero variance by letting $M \to \infty$, because of the nonindependence of the $b_{ikm}^*$.

## 5.4 ESTIMATED ALLOCATION RATES FOR HEMOPHILIA DATA

Basford and McLachlan (1985a) considered the assessments in terms of the estimated allocation rates of the clusterings produced by the mixture likelihood approach in its application to three real data sets. We report here their results for one of these sets, namely the hemophilia data which were described in Section 3.2. For this set there are $n = 75$ bivariate observations of which $n_1 = 30$ are on known noncarriers (population $G_1$) and $n_2 = 45$ are on known carriers (population $G_2$). It was observed there (Figure 4) that in fitting a mixture $G$ of two bivariate normal distributions with unequal variances, the data were clustered into two groups corresponding to $G_1$ and $G_2$, with correct allocation rates given by $A_1 = 27/30$ and $A_2 = 33/45$ respectively. The overall correct rate was therefore $A = 60/75$.

The estimates of these allocation rates as given by $T_1$, $T_2$ and $T$ are displayed in Table 19, along with the bootstrap estimates of their root mean squared errors and biases. The standard errors of the latter Monte Carlo approximations, $\bar{b}_1^*$, $\bar{b}_2^*$, and $\bar{b}^*$, to the bootstrap biases, $b_1$, $b_2$, and $b$, are given in parentheses. The bias corrected estimates of the allocation rates obtained as a consequence, $T_1^{(B)}$, $T_2^{(B)}$, and $T^{(B)}$, are also displayed in Table 19.

It can be seen from this table that the bootstrap method is effective in reducing the bias of not only the overall allocation rate but also each individual population rate. For instance, bias correction of the overall estimated rate reduced the distance between the initial estimate and the true rate by a factor of 63%. Contrasting the results for the semiparametric version of the bootstrap method with those of the parametric version, it can be seen that for this example the latter is clearly preferable.

## 5.5 ESTIMATED ALLOCATION RATES FOR SIMULATED DATA

As mentioned previously, Basford and McLachlan (1985a) carried out some

**TABLE 19** Estimation of correct allocation rates for hemophilia data set using bootstrap method of assessment of bias and root mean squared error

| Source | True rate | Estimated rate | Estimate of RMSE | Estimate of bias | (s.e.) | Corrected estimate |
|---|---|---|---|---|---|---|
| | | | Parametric version | | | |
| $G_1$ | 0.900 | 0.938 | 0.113 | 0.052 | (0.017) | 0.886 |
| $G_2$ | 0.733 | 0.902 | 0.193 | 0.090 | (0.024) | 0.812 |
| $G$ | 0.800 | 0.921 | 0.120 | 0.076 | (0.013) | 0.845 |
| | | | Semiparametric version | | | |
| $G_1$ | 0.900 | 0.938 | 0.052 | 0.009 | (0.007) | 0.929 |
| $G_2$ | 0.733 | 0.902 | 0.133 | 0.060 | (0.017) | 0.842 |
| $G$ | 0.800 | 0.921 | 0.071 | 0.035 | (0.009) | 0.883 |

Adapted from Basford and McLachlan (1985a)

simulations in order to investigate the usefulness of the estimated allocation rates and their bias corrected versions, as given by the bootstrap, in assessing the strength of the clustering produced by the mixture approach. Four different specifications, labeled Cases A, B, C, and D, were considered for a mixture $G$ in equal proportions of two bivariate normal populations $G_1$ and $G_2$ with different means $\boldsymbol{\mu}_1$ and $\boldsymbol{\mu}_2$ and common covariance matrix $\Sigma$ (see Table 18). Without loss of generality, it was assumed that $\Sigma$ was the identity matrix and that

$$\boldsymbol{\mu}_1 = (\Delta, 0)' \qquad \text{and} \qquad \boldsymbol{\mu}_2 = (0, 0)',$$

where $\Delta$ is the Mahalanobis distance between $G_1$ and $G_2$.

The clustering of the simulated normal data sets with the mixture likelihood approach was based on the solution of the likelihood equation for $\phi$ corresponding to the application of the EM algorithm started from $\hat{\phi}_C$, the maximum likelihood estimate of $\phi$ formed with knowledge of the true origin of the data. This information is not available in practice, so results obtained in a practical setting may not always be as favorable.

Before we give a summary of their overall conclusions for the four sets, we focus briefly on Case A to illustrate the scope of their simulation experiments. Case A is the least favorable for clustering in terms of design since the Mahalanobis distance $\Delta$ between the two populations is the smallest ($\Delta = \sqrt{2}$). In order to demonstrate the formation of the bootstrap estimates of the bias and mean squared error of the estimated allocation rates simulated in Case A, Basford and McLachlan (1985a) tabulated the values of $T^*$ and $A^*$ for the first five bootstrap replications performed on the first of the 40 trials simulated in Case A. These values are displayed here in Table 20. For this trial $T = 0.854$ and $A = 0.767$. The bootstrap estimate of the bias of $T$ is given by $\bar{b}^*$, obtained by averaging the values of $b^* = T^* - A^*$ from the $K = 50$ bootstrap replications performed. For this trial, $\bar{b}^* = 0.115$, leading to $T^{(B)} = T - \bar{b}^* = 0.739$ as the bias corrected estimate of $A$. The root mean squared error of $T$ was estimated by the positive square root of the average over the $K = 50$ bootstrap replications of $b^{*2}$, yielding $\sqrt{0.0226} = 0.150$ as the estimate.

The above exercise was repeated for each of 40 simulation trials, and the results obtained are given in Table 21; the standard errors are in parentheses. In this table, RMSE and RMSE$^{(B)}$ refer to the root mean squared errors of the estimated rate and its bias corrected version, assessed by the positive square root of the averages over the simulated values of the appropriate quantities, for example, $(T - A)^2$ and $(T^{(B)} - A)^2$, respectively, for the overall allocation rate. Further, RMSE$^{(IC)}$ refers to the root mean squared error of the "ideal constant" estimated rate, but with the unknown

**TABLE 20** The first five bootstrap replications and summary statistics for $K = 50$ bootstrap replications for the overall allocation rate on Trial 1 in Case A of simulation study for which $T = 0.854$ and $A = 0.767$

| Bootstrap replication | $T^*$ | $A^*$ | $b^* = T^* - A^*$ | $b^{*2}$ |
|---|---|---|---|---|
| 1 | 0.771 | 0.700 | 0.071 | 0.005 |
| 2 | 0.962 | 0.817 | 0.145 | 0.021 |
| 3 | 0.888 | 0.800 | 0.088 | 0.008 |
| 4 | 0.907 | 0.833 | 0.074 | 0.005 |
| 5 | 0.922 | 0.733 | 0.189 | 0.036 |
| 50 replications: Average | | | 0.115 | 0.023 |

Source: Basford and McLachlan (1985a)

bias replaced by its simulated value. For example, in forming

$$T^{(\mathrm{IC})} = T - \beta,$$

the bias $\beta$ was replaced by $\overline{T} - \bar{A}$, the value of $T - A$ averaged over the 40 simulation trials. Also listed in Table 21 is CORR, the correlation between the (uncorrected) estimated rate minus the actual rate and the bootstrap estimator of the expectation of this difference. Note that since the estimated overall allocation rate, $T$, can be expressed as

$$T = \hat{\pi}_1 T_1 + \cdots + \hat{\pi}_g T_g,$$

it lies within the range of the estimates of the individual rates $T_i$ ($i = 1, \ldots, g$). This does not necessarily apply, however, to the rates after they have been averaged over the number of simulation trials performed in each case. For example, in Case A, it can be seen that $\overline{T} = 0.840$ lies just outside the range of the means of the estimates of the individual rates, given by $\overline{T}_1 = 0.839$ and $\overline{T}_2 = 0.802$.

Encouraging results were obtained in each of the four cases of the simulation study by Basford and McLachlan (1985a), suggesting that the estimation of the allocation rates and their subsequent bias correction according to the bootstrap is a very worthwhile exercise. In each of the four cases, the bootstrap method of bias correction appreciably reduced the optimistic assessment provided in the first place by the uncorrected estimates. Moreover, bias correction of the estimated rates reduced the root mean squared

**TABLE 21** Summary results for Case A of simulation study

| | $G_1$ | $G_2$ | $G$ |
|---|---|---|---|
| True rate | 0.764 | 0.697 | 0.740 |
| | (0.024) | (0.031) | (0.010) |
| Estimated rate | 0.839 | 0.802 | 0.840 |
| | (0.021) | (0.026) | (0.014) |
| Corrected estimate | 0.767 | 0.731 | 0.762 |
| | (0.024) | (0.028) | (0.019) |
| Estimate of RMSE | 0.137 | 0.146 | 0.117 |
| | (0.008) | (0.007) | (0.007) |
| RMSE | 0.146 | 0.194 | 0.154 |
| $\text{RMSE}^{(B)}$ | 0.134 | 0.193 | 0.147 |
| $\text{RMSE}^{(IC)}$ | 0.124 | 0.163 | 0.118 |
| CORR | −0.163 | −0.590 | −0.625 |

Adapted from Basford and McLachlan (1985a).

error of each estimated rate in every case. A comparison of the root mean squared error of the bias corrected estimated rate with that of the "ideal constant" estimated rate reveals the progress made in reducing the mean squared error to this level.

The correlation between $T - A$ and $\bar{b}^*$ was simulated to be negative in Cases A, B, and C, while in Case D, it was positive but very close to zero. For the individual allocation rates the corresponding simulated correlations were negative in Cases A and C, but of smaller magnitude than for the overall rate. The presence of a negative correlation between the estimated allocation rate minus the actual rate and the bootstrap estimator of the bias inflates the mean squared error of the bias corrected version of the estimated rate. For example, consider the mean squared error of the estimated overall allocation rate corrected for bias according to the bootstrap. It is given by

$$\begin{aligned}\text{MSE}(T^{(B)}) &= E\{T^{(B)} - A\}^2 \\ &= E(T - \bar{b}^* - A)^2,\end{aligned}$$

which can be expressed as

$$\mathrm{MSE}(T^{(B)}) = \mathrm{var}(T - A) + \mathrm{var}(\bar{b}^*) + \left\{E(\bar{b}^*) - \beta\right\}^2 - 2\,\mathrm{cov}(T - A, \bar{b}^*).$$

It can be seen that a negative value for the correlation between $T - A$ and $\bar{b}^*$ increases the mean squared error of $T^{(B)}$. With the bootstrap method of bias correction of the apparent error rate in a discriminant analysis context, Efron (1982) noted a marked negative correlation between the apparent error rate minus the actual rate and the bootstrap estimator of the bias. Efron (1983) subsequently developed for that problem more sophisticated variants of his bootstrap, which were found to reduce the negative correlation nearly to zero without increasing the variance of the estimated bias much above that for the ordinary bootstrap. Concerning the performance of the bootstrap method in estimating the root mean squared errors of the estimated individual and overall allocation rates in the four cases considered by Basford and McLachlan (1985a), there was generally good agreement between the estimated and simulated root mean squared errors.

## 5.6 OTHER METHODS OF BIAS CORRECTION

Other methods, besides the bootstrap, for estimating bias and in some instances variance include the jackknife, the delta method and cross-validation. Efron (1982) gave an excellent account of the available methods. In particular, he examined the relationships between the methods, and identified those situations in which estimates produced according to some of the methods are similar or indeed the same.

The nonparametric jackknife procedure, which sequentially deletes observations and recomputes the estimates, is a commonly used method. It was introduced by Quenouille (1949, 1956) for estimating bias, and later utilized by Tukey (1958) to provide a nonparametric estimate of variance; see Miller (1974) for a comprehensive review of this topic. Cross-validation relates to the situation where one has to make a prediction about a new observation, and is a way of reducing the bias caused by assessing the performance of the prediction rule on the basis of the same sample used to form it. Efron (1982) showed in the regression context that the jackknife, cross-validation and the bootstrap are closely connected in theory, if not necessarily in their practical consequences. Concerning the concept of excess error in this context, it was shown that the jackknife approximation to the bootstrap estimate is very similar to the cross-validated estimate. On the same concept, McLachlan (1980b) performed some limited simulations to illustrate the efficiency of the bootstrap relative to the parametric delta method.

These competitors of the bootstrap, however, appear not to be applicable in a straightforward manner to the current problem where the aim is to reduce the biases of the $T_i$ and $T$ as estimators of the unobservable allocation rates $A_i$ and $A$. For instance, application of the delta method would require the derivation of the asymptotic expansions of $E(T_i - A_i)$ and $E(T - A)$, which would involve tedious calculations even if they were tractable.

It is of interest to consider the application of the jackknife to this problem. It obviously cannot be applied to the estimates themselves. Consider the estimation of $T$; the jackknife produces an estimate by extrapolating $T$ to its value at $n = \infty$, $T_\infty$. The requirement, however, is an unbiased estimator of $E(A)$, not of $T_\infty$, and the former may not be close to $T_\infty$ for small $n$, although it does equal $T_\infty$ in the limit. Moreover, there does not appear to be a useful role for the jackknife in providing approximations to the bootstrap expectations of the statistics $T_i^* - A_i^*$ and $T^* - A^*$. Efron (1982) showed how the jackknife can be used in certain situations to approximate bootstrap expectations and variances, and so reduce their computation. For example suppose that the vector of indicator variables, $\mathbf{z}_j = (\mathbf{z}'_{1j}, \ldots, \mathbf{z}'_{gj})'$ associated with each $\mathbf{x}_j$ in the initial sample $\mathbf{x}_1, \ldots, \mathbf{x}_n$ were known; the rates $A_i$ and $A$ would also be known then. Suppose further, that for some reason, it is desired to estimate the expectation of $T - A$, where $T$ is still to be formed as before using estimates computed under the mixture model without knowledge of the $\mathbf{z}_j$, but where the known values of the $\mathbf{z}_j$ are to be used in forming $A$ in the subsequent estimation of $E(T - A)$. It follows from Efron (1982) that the jackknife-type expression,

$$H^\circ + (n-1)(\bar{H} - H^\circ), \tag{5.6.1}$$

can be considered as an approximation to the bootstrap expectation of $T^* - A^*$, where $H^\circ$ denotes the value of $H = T - A$ based on $\mathbf{x}_1, \ldots, \mathbf{x}_n$ and $H_{(i)}$ denotes its value with $\mathbf{x}_i$ deleted, and

$$\bar{H} = \sum_{i=1}^{n} H_{(i)}/n.$$

This is because (5.6.1) is equal to $n/(n-1)$ times the bootstrap expectation (with respect to the empirical distribution function) of any quadratic function agreeing with $H$ at the $n$-dimensional points $\mathbf{Q}^\circ$, $\mathbf{Q}_{(1)}, \ldots, \mathbf{Q}_{(n)}$, where $(\mathbf{Q}^\circ)_j = 1/n$ $(j = 1, \ldots, n)$ and $(\mathbf{Q}_{(i)})_j$ is zero for $j = i$ and $1/(n-1)$ otherwise. The random variable $H$ is regarded as a functional $H(\hat{F}(\mathbf{Q}))$ with $\hat{F}(\mathbf{Q})$ denoting the distribution function which attaches weight $(\mathbf{Q})_j$ to $\mathbf{x}_j$, where the $\mathbf{x}_j$ are fixed; $\hat{F}(\mathbf{Q}^\circ)$, for example, is the empirical distribution function. If when considered as a functional, $H$ were also a function of the unknown distribution function $F$, $H = H(\hat{F}, F)$, then the jackknife approximation to the bootstrap expectation of $H(\hat{F}(\mathbf{Q}^*), \hat{F}(\mathbf{Q}^\circ))$ in a nonparametric framework is formed simply with $F$ fixed at $\hat{F}(\mathbf{Q}^\circ)$ in $H$; $\mathbf{Q}^*$ denotes the bootstrap analogue of $\mathbf{Q}$. The complication with the nonparametric jackknife in the actual problem at hand is that the unknown quantities $\mathbf{z}_j$ on which $A$, and hence $H$, depend are random variables and not parameters. It has been shown that the bootstrap method overcomes this obstacle by adopting a parametric approach. For example, with the semiparametric version of the bootstrap, where the bootstrap sample $\mathbf{x}_1^*, \ldots, \mathbf{x}_n^*$ is obtained nonparametrically, the corresponding bootstrap labels $\mathbf{z}_1^*, \ldots, \mathbf{z}_n^*$ are subsequently generated by using the parametric forms for the posterior probabilities of the $\mathbf{x}_j^*$. In any event, there appears little to be gained here by seeking a jackknife-type approximation. There would be no saving in computation as the parametric bootstrap expectations can be fairly well estimated by the Monte Carlo method using, say 50, bootstrap replications. Moreover, Efron (1982) demonstrated for various other problems, including estimation of the excess error rate in discriminant analysis, that the jackknife gives a more variable estimate than the bootstrap Monte Carlo approximation.

Overall, it can be seen that the estimation problem considered here is far from straightforward. Nevertheless, it can be handled directly by either a parametric or semiparametric application of the bootstrap method.

## 5.7 BIAS CORRECTION FOR ESTIMATED POSTERIOR PROBABILITIES

Unless $n$ is large, the bias of the likelihood estimator of $\phi$ under the fitted mixture model will not be small, and this bias will manifest itself in the estimates of the posterior probabilities. Thus consideration might be given to the use of a bias corrected version of $\hat{\phi}$ in forming the estimated posterior probabilities, with a view to improving the allocation of the sample. For normal mixtures bias correction of $\hat{\phi}$ using the bootstrap method would involve term by term additive correction of a covariance matrix, and this is fraught with danger. This can be avoided, however, by writing the estimated posterior probability that $\mathbf{x}$ belongs to population $G_i$ in the form

$$\tau_i(\mathbf{x};\hat{\phi}) = \varrho_i(\mathbf{x};\hat{\phi}) \Bigg/ \left\{1 + \sum_{t=1}^{g-1} \varrho_t(\mathbf{x};\hat{\phi})\right\} \qquad (i = 1, \ldots, g-1),$$

where

$$\varrho_i(\mathbf{x};\hat{\phi}) = \exp(\mathbf{x}'\hat{\boldsymbol{\kappa}}_{1i}\mathbf{x} + \mathbf{x}'\hat{\boldsymbol{\kappa}}_{2i} + \hat{\kappa}_{3i}) \qquad (i = 1, \ldots, g-1)$$

and where

$$\hat{\boldsymbol{\kappa}}_{1i} = \tfrac{1}{2}(\hat{\boldsymbol{\Sigma}}_g^{-1} - \hat{\boldsymbol{\Sigma}}_i^{-1}),$$

$$\hat{\boldsymbol{\kappa}}_{2i} = \hat{\boldsymbol{\Sigma}}_i^{-1}\hat{\boldsymbol{\mu}}_i - \hat{\boldsymbol{\Sigma}}_g^{-1}\hat{\boldsymbol{\mu}}_g,$$

and

$$\hat{\kappa}_{3i} = \log(\hat{\pi}_i/\hat{\pi}_g) + \tfrac{1}{2}\log\{|\hat{\boldsymbol{\Sigma}}_g|/|\hat{\boldsymbol{\Sigma}}_i|\} + \tfrac{1}{2}(\hat{\boldsymbol{\mu}}_g'\hat{\boldsymbol{\Sigma}}_g^{-1}\hat{\boldsymbol{\mu}}_g - \hat{\boldsymbol{\mu}}_i'\hat{\boldsymbol{\Sigma}}_i^{-1}\hat{\boldsymbol{\mu}}_i).$$

As in the previous work, $\boldsymbol{\mu}_i$ and $\boldsymbol{\Sigma}_i$ denote the mean vector and covariance matrix of the normal density of an observation in $G_i$ ($i = 1, \ldots, g$). Bias correction can then be undertaken by applying the bootstrap method directly to the $\hat{\boldsymbol{\kappa}}_{1i}$, $\hat{\boldsymbol{\kappa}}_{2i}$, and the $\hat{\kappa}_{3i}$ instead of to the $\hat{\boldsymbol{\Sigma}}_i$, $\hat{\boldsymbol{\mu}}_i$, and the $\hat{\pi}_i$. As a consequence of the bias correction to $\hat{\phi}$ the relative sizes of the posterior probabilities of observations of doubtful population membership may change sufficiently so that some of them may be put in different groups in an outright assignment of the sample in terms of the estimated posterior probabilities.

Bias correction of the likelihood estimate of $\phi$ using the bootstrap method was carried out by Basford (1985) for the hemophilia data set just considered in Section 5.4 and initially described in Section 3.2. There the likelihood estimate of $\phi$ under a mixture model consisting of two bivariate normal, heteroscedastic populations was denoted by $\hat{\phi}_1$. However, in this section, the subscript denoting the particular local maximum will be deleted, and so this estimate will be denoted by $\hat{\phi}$. The bias corrected version of $\hat{\phi}$ obtained using a fully parametric form of the bootstrap is denoted by $\hat{\phi}_B$ and listed in Table 22, along with $\hat{\phi}$ and $\hat{\phi}_C$, the estimate of $\hat{\phi}$ based on the true origin of the data.

The estimated posterior probability of belonging to $G_1$ (population of noncarriers) formed before and after bias correction of $\hat{\phi}$ is listed in Table 23 for each of the 30 known noncarriers $\mathbf{x}_{1j}$ ($j = 1, \ldots, 30$) and for each of the 45 known carriers $\mathbf{x}_{2j}$ ($j = 1, \ldots, 45$). Concerning the observations

**TABLE 22** Estimates of $\phi$ for hemophilia data set

| $\phi =$ | $(\pi_1,$ | $\pi_2,$ | $(\boldsymbol{\mu}_1)_1,$ | $(\boldsymbol{\mu}_1)_2,$ | $(\boldsymbol{\mu}_2)_1,$ | $(\boldsymbol{\mu}_2)_2$ | $(\Sigma_1)_{11}$ | $(\Sigma_1)_{12}$ | $(\Sigma_1)_{22}$ | $(\Sigma_2)_{11}$ | $(\Sigma_2)_{12}$ | $(\Sigma_2)_{22})'$ |
|---|---|---|---|---|---|---|---|---|---|---|---|---|
| $\hat{\phi} =$ | (0.503, | 0.497, | −11.4, | −2.4, | −36.4, | −4.5, | 111, | 65, | 123, | 160, | 150, | 321)′ |
| $\hat{\phi}_B =$ | (0.483, | 0.517, | −9.8, | −2.0, | −36.0, | −4.5, | 100, | 70, | 130, | 160, | 160, | 360)′ |
| $\hat{\phi}_C =$ | (0.400, | 0.600, | −13.5, | −7.8, | −30.8, | −0.6, | 209, | 155, | 179, | 238, | 154, | 240)′ |

$\mathbf{x}_{1j}$ from $G_1$, all the $\tau_1(\mathbf{x}_{1j}; \hat{\phi})$ are near to 0 or 1, and bias correction does not change the estimates to any marked degree, with the exception perhaps of $j = 7$ for which membership of $G_1$ becomes more doubtful. For some of the observations from $G_2$, however, their estimated posterior probabilities of belonging to this population are near 0.5 and bias correction of $\phi$ does result in a different allocation. It can be seen that the observations $\mathbf{x}_{2j}$ from $G_2$ with $j = 6$, 7, and 9 are now correctly allocated as their estimated posterior probability of belonging to $G_1$ changes from above 0.5 to slightly below 0.5. The observation $\mathbf{x}_{2j}$ with $j = 39$ is still misallocated to $G_1$, but its estimated posterior probability of belonging to $G_1$ is so close to 0.5 that it is now in the extremely doubtful category. For this example, there does not appear to be any clear pattern as to whether the estimated posterior probabilities are more or less extreme as a consequence of bias correction.

**TABLE 23** Estimated posterior probability of being a noncarrier for 30 known noncarriers $\mathbf{x}_{1j}$ ($j = 1, \ldots,$ 30) and 45 known carriers $\mathbf{x}_{2j}$ ($j = 1, \ldots, 45$) before and after bias correction of $\hat{\phi}$

| | Noncarriers | | Carriers | |
|---|---|---|---|---|
| $j$ | $\tau_1(\mathbf{x}_{1j}; \hat{\phi})$ | $\tau_1(\mathbf{x}_{1j}; \hat{\phi}_B)$ | $\tau_1(\mathbf{x}_{2j}; \hat{\phi})$ | $\tau_1(\mathbf{x}_{2j}; \hat{\phi}_B)$ |
| 1 | 1.00 | 1.00 | 0.00 | 0.00 |
| 2 | 0.97 | 0.97 | 0.00 | 0.00 |
| 3 | 0.23 | 0.12 | 0.94 | 0.94 |
| 4 | 0.98 | 0.98 | 0.21 | 0.10 |
| 5 | 0.69 | 0.62 | 0.12 | 0.04 |
| 6 | 0.99 | 0.98 | 0.59 | 0.45 |
| 7 | 0.76 | 0.69 | 0.60 | 0.48 |
| 8 | 0.98 | 0.98 | 0.72 | 0.63 |
| 9 | 0.95 | 0.95 | 0.57 | 0.43 |
| 10 | 0.99 | 0.99 | 0.14 | 0.07 |
| 11 | 0.00 | 0.00 | 0.89 | 0.89 |
| 12 | 0.98 | 0.98 | 0.23 | 0.11 |
| 13 | 1.00 | 1.00 | 0.97 | 0.97 |
| 14 | 1.00 | 1.00 | 0.97 | 0.97 |
| 15 | 0.94 | 0.94 | 0.94 | 0.93 |
| 16 | 1.00 | 1.00 | 0.96 | 0.96 |
| 17 | 0.01 | 0.00 | 0.99 | 0.99 |
| 18 | 0.96 | 0.96 | 0.07 | 0.02 |
| 19 | 1.00 | 1.00 | 0.00 | 0.00 |
| 20 | 0.85 | 0.84 | 0.17 | 0.08 |

**Table 23** *Continued*

| | Noncarriers | | Carriers | |
|---|---|---|---|---|
| $j$ | $\tau_1(\mathbf{x}_{1j};\hat{\phi})$ | $\tau_1(\mathbf{x}_{1j};\hat{\phi}_B)$ | $\tau_1(\mathbf{x}_{2j};\hat{\phi})$ | $\tau_1(\mathbf{x}_{2j};\hat{\phi}_B)$ |
| 21 | 0.95 | 0.85 | 0.00 | 0.00 |
| 22 | 1.00 | 1.00 | 0.00 | 0.00 |
| 23 | 0.98 | 0.98 | 0.01 | 0.00 |
| 24 | 0.97 | 0.97 | 0.01 | 0.00 |
| 25 | 0.92 | 0.90 | 0.00 | 0.00 |
| 26 | 1.00 | 1.00 | 0.04 | 0.00 |
| 27 | 0.89 | 0.87 | 0.00 | 0.00 |
| 28 | 0.99 | 0.99 | 0.01 | 0.00 |
| 29 | 0.88 | 0.84 | 0.26 | 0.13 |
| 30 | 0.99 | 0.99 | 0.00 | 0.00 |
| 31 | | | 0.00 | 0.00 |
| 32 | | | 0.00 | 0.00 |
| 33 | | | 0.08 | 0.05 |
| 34 | | | 0.01 | 0.00 |
| 35 | | | 0.00 | 0.00 |
| 36 | | | 0.03 | 0.01 |
| 37 | | | 0.03 | 0.01 |
| 38 | | | 0.41 | 0.32 |
| 39 | | | 0.62 | 0.54 |
| 40 | | | 0.08 | 0.03 |
| 41 | | | 0.01 | 0.00 |
| 42 | | | 0.01 | 0.01 |
| 43 | | | 0.12 | 0.06 |
| 44 | | | 0.03 | 0.01 |
| 45 | | | 0.00 | 0.00 |

Basford and McLachlan (1985a) concluded from their empirical investigation that the bias corrected versions of the estimated allocation rates $T_i$ and $T$ still give an optimistic assessment of the true rates for the mixture approach based on $\hat{\phi}$. Whether these bias corrected estimated rates provide an optimistic assessment of the true rates for the mixture approach based on $\hat{\phi}_B$ is not clear. This is just one of the problems concerned with the use of $\hat{\phi}_B$ which is in need of future research. It might be anticipated that the use of $\hat{\phi}_B$ would improve the performance of the mixture approach, so that its correct allocation rates would be increased with respect to some or all of the populations. If this is the case, then the bias corrected versions of $T_i$ and $T$ as obtained for the mixture approach based on $\hat{\phi}$ will provide a less optimistic assessment of the mixture approach based on $\hat{\phi}_B$. Actually, this does happen in the example taken here. From Table 19, the true allocation rates for the mixture approach based on $\hat{\phi}$ are given by $A_1 = 0.9$, $A_2 = 0.733$ and $A = 0.8$, and it has just been seen that the use of $\hat{\phi}_B$ gives an improved allocation of the sample with $A_2$, and hence $A$, being increased to 0.8 and 0.84, respectively. Now the bias corrected versions of $T_1$, $T_2$ and $T$, as estimates of the allocation rates of the mixture approach based on $\hat{\phi}$, are equal to 0.886, 0.812 and 0.845, respectively, and so the bias corrected versions of $T_2$ and $T$ are actually closer to the true rates for the mixture approach on $\hat{\phi}_B$.

Of course it is possible to obtain bias corrected versions of the estimated allocation rates of the mixture approach using $\hat{\phi}_B$, but it would involve two layers of bootstrapping; that is, $K^2$ replications which would be 2,500 recomputations here. Note that even if the $T_i$ and $T$ were formed using unbiased estimators of the posterior probabilities, they would still be biased estimators of the corresponding allocation rates. Although the results discussed in this section refer to a single empirical example, they do indicate that bias correction of $\hat{\phi}$ is worthy of further research.

# 6
# Partitioning of Treatment Means in ANOVA

## 6.1 INTRODUCTION

A common problem in practice is the analysis of experiments designed to compare several treatments. In the situation where an analysis of variance leads to a significant $F$ test for the difference between treatments there is the problem of isolating those treatments which do not appear to be different from each other. Tukey (1949) succinctly described this in a hypothetical example when he said, "At a low and practical level, what do we want to do? We wish to separate the varieties into distinguishable groups as often as we can without too frequently separating varieties which should stay together."

One of the early attempts at this problem of assessing where the real differences lie was the simple method of comparing all differences between the treatment sample means with the least significant differences (LSD), using Student's $t$ distribution. Various multiple comparison procedures have since been suggested, and a survey of work in this area may be found in the book by Miller (1981) and the review article by O'Neill and Wetherill (1971). A recent account of some nonparametric approaches to multiple comparisons in a one-way analysis of variance has been given by Skillings (1983). As noted by O'Neill and Wetherill (1971), there has been much confusion in the area of multiple comparisons, for example, over what the basic problems really are and what the various procedures should achieve. Indeed, some statisticians seem to be opposed to such methods.

Notable among multiple comparison procedures are the Studentized range methods due primarily to Tukey (1949) and the $F$ test procedure of Gabriel (1964). These procedures lead to an underscoring of those subsets of the ordered sample means which are judged to be not significantly different among themselves. However, as emphasized by Caliński and Corsten (1985), the overlapping of the homogeneous subsets produced by multiple comparison procedures such as the two mentioned above is regarded as a disadvantage by many statistical users, for example, plant breeders. The primary aim therefore in this context would appear to be to partition the treatments into nonoverlapping groups so that treatments within a group can be considered to be fairly homogeneous. Ideally, there should be good separation between the groups.

A partition of the treatments with these aims in mind can be attempted using a cluster analysis method. The possibility of using cluster analysis in place of a multiple comparison procedure, had been suggested by Plackett (1971) in his discussion of the paper by O'Neill and Wetherill (1971). Scott and Knott (1974) used a hierarchical, divisive cluster analysis method to split the treatment means from a one-way experimental design into homogeneous groups. Also, Jolliffe (1975) applied hierarchical clustering procedures to this problem. Aitkin (1980a) clustered treatment means from a one-way design by fitting a normal mixture model. More recently, Basford and McLachlan (1985b) considered the fitting of a normal mixture to data from a randomized complete block design with both fixed and random block effects. Binder (1978, 1981), who developed a Bayesian approach to cluster analysis, applied his technique to data on yields from seven varieties of barley in a randomized complete block design. This data set, which shall be considered later in this chapter using a number of clustering methods, was taken from Duncan (1955), one of the early papers on multiple comparisons. Menzefricke (1981) extended Binder's (1978) Bayesian clustering technique to the one-way multivariate analysis of variance.

Additional references on this problem may be found in Carmer and Lin (1983), who compared five univariate, divisive clustering methods for grouping means in an analysis of variance. Caliński and Corsten (1985) considered the grouping of treatment means using two clustering methods which were embedded in a consistent manner into simultaneous test procedures. One of their methods is implemented hierarchically on the basis of the Studentized range, while the other proceeds in a nonhierarchical manner in terms of the pooled within group sum of squares, which was used by Scott and Knott (1974) in their hierarchical clustering method. Also, Cox and Spjøtvoll (1982) used this sum of squares to give a simple method based on standard $F$ tests for partitioning means into nonoverlapping groups. However, their proposal differs from the clustering type methods in that it does not attempt to specify one partition for each level of $g$. Rather its conclusions concern all partitions which are compatible with the data for the particular $g$. These methods are to be described in more detail in Section 6.4.

In this chapter we shall be focusing on the role of mixture models with normal component distributions for the clustering of treatment means. In the discussion of the paper by O'Neill and Wetherill (1971), Nelder (1971) commented that one of the patterns to look for in the sample means of the treatments was whether "the means divide into two or more groups within which they look like samples from normal distributions." It can be seen that the fitting of a normal mixture to the yields of the treatments is in keeping with the spirit of this comment.

It is assumed that there are $n$ treatments with unknown means $\mu_1, \ldots, \mu_n$. Rather than specify a particular experimental design from which the data on these treatments have been obtained, it is assumed in the first instance that there are available $n$ independent sample means $\bar{x}_1, \ldots, \bar{x}_n$ which are normally distributed with means $\mu_1, \ldots, \mu_n$ and variances $\sigma^2/r_1, \ldots, \sigma^2/r_n$ respectively, and an independent estimate $s^2$ of $\sigma^2$, distributed as

$$\nu s^2/\sigma^2 \sim \chi^2_\nu.$$

This covers various experimental designs, including a completely randomized design with unequal replicates and, for $r_j = r$ $(j = 1, \ldots, n)$, more involved designs with equireplicated treatments such as a Latin square experiment or a randomized complete block design, where there are present other main effects orthogonal to treatments, for example, block effects in the latter design.

As explained above, the aim is to divide the $n$ treatment means into $g$ $(\leq n)$ distinct groups $G_1, \ldots, G_g$, so that within each group the treatment means are the same. We relabel the treatment means so that $\mu_1, \ldots, \mu_g$ now denote the distinct values, with $\mu_i$ the common value of the treatment means in $G_i$ $(i = 1, \ldots, g)$. We let $\mathbf{z}_j = (z_{1j}, \ldots, z_{gj})'$ be the vector of indicator variables denoting group membership for treatment $j$, where $z_{ij} = 1$ if treatment $j$ (more precisely, if the mean of treatment $j$) belongs to $G_i$, and 0 otherwise. By fitting a mixture model, an assessment of the group membership of the $n$ treatment means as represented by $\mathbf{Z} = (\mathbf{z}'_1, \ldots, \mathbf{z}'_n)'$ is obtained in terms of the estimated expectation of $\mathbf{Z}$ conditional on the available data; that is, in terms of the estimated posterior probabilities of group membership for the treatments. These can be easily calculated for a completely randomized design or for a randomized complete block design with fixed block effects but, for random block effects, Basford and

McLachlan (1985b) have shown that the amount of computation becomes prohibitive for a moderate number of treatments. A guide to the number of groups, $g$, is provided by the likelihood ratio test.

The mixture likelihood approach to this problem shall be demonstrated on a number of real data sets and the results contrasted with some other analyses as reported in the literature.

## 6.2 CLUSTERING OF TREATMENT MEANS BY THE MIXTURE LIKELIHOOD APPROACH

We consider the mixture likelihood approach to the clustering of treatment means firstly in the situation where the available data on the $n$ treatment means consist of $n$ independent observations $\bar{x}_1, \ldots, \bar{x}_n$ distributed normally about these means with variances $\sigma^2/r_1, \ldots, \sigma^2/r_n$ respectively, and where there is an independent estimate of $\sigma^2$, $s^2$, distributed as

$$\nu s^2/\sigma^2 \sim \chi^2_\nu. \tag{6.2.1}$$

Under a normal mixture model with $g$ components, it is assumed that

$$\bar{x}_j \sim N(\mu_i, \sigma^2/r_j) \qquad \text{in } G_i \text{ with prob. } \pi_i \quad (i = 1, \ldots, g) \tag{6.2.2}$$

for $j = 1, \ldots, n$. We let

$$\phi = (\pi', \mu_1, \ldots, \mu_g, \sigma^2)'$$

be the vector of unknown parameters and $\phi_j$ denote $\phi$ with $\sigma^2$ replaced by $\sigma^2/r_j$ ($j = 1, \ldots, n$). The log likelihood function of $\phi$ formed on the basis of $\bar{x}_1, \ldots, \bar{x}_n$ under the mixture model (6.2.2) and also $s^2$ under (6.2.1) is given, up to terms involving $\phi$, by

$$L(\phi) = \sum_{j=1}^{n} \log f(\bar{x}_j; \phi_j) - \tfrac{1}{2}\nu \log \sigma^2 - \tfrac{1}{2}\nu s^2/\sigma^2$$

where $f(x; \phi_j)$ denotes a mixture of $g$ univariate normal densities with means $\mu_1, \ldots, \mu_g$ and common variance $\sigma^2/r_j$ ($j = 1, \ldots, n$). From the assumption of constant variance in the analysis of variance, each $\sqrt{r_j}\bar{x}_j$ has the same variance $\sigma^2$ in each group. Hence the maximum likelihood estimator of $\phi$ exists and, from the theory given in Section 2.2, is consistent and asymptotically efficient.

The log likelihood function for the complete data, which contains also **Z** specifying the indicator variables $z_{ij}$ of group membership for the treatments, is given, up to terms involving $\phi$, by

$$L_C(\phi) = \sum_{i=1}^{g}\sum_{j=1}^{n} z_{ij}\left\{\log \pi_i + \log f(\bar{x}_j; \mu_i, \sigma^2/r_j)\right\}$$

$$- \tfrac{1}{2}\nu \log \sigma^2 - \tfrac{1}{2}\nu s^2/\sigma^2,$$

where $f(x;\ \mu_i,\ \sigma^2/r_j)$ denotes a univariate normal density with mean $\mu_i$ and variance $\sigma^2/r_j$ $(i = 1, \ldots, g;\ j = 1, \ldots, n)$. By consideration of $L_C(\phi)$ under the application of the EM algorithm, it follows that $\hat{\phi}$, the maximum likelihood estimate of $\phi$ under (6.2.1) and the mixture model (6.2.2), satisfies

$$\hat{\pi}_i = \sum_{j=1}^{n} \hat{\tau}_{ij}/n \qquad (i = 1, \ldots, g), \tag{6.2.3}$$

$$\hat{\mu}_i = \sum_{j=1}^{n} \hat{\tau}_{ij} r_j \bar{x}_j \bigg/ \sum_{j=1}^{n} \hat{\tau}_{ij} r_j \qquad (i = 1, \ldots, g), \tag{6.2.4}$$

and

$$\hat{\sigma}^2 = \left\{\nu s^2 + \sum_{i=1}^{g}\sum_{j=1}^{n} \hat{\tau}_{ij} r_j (\bar{x}_j - \hat{\mu}_i)^2\right\} \bigg/ (\nu + n), \tag{6.2.5}$$

where

$$\hat{\tau}_{ij} = \pi_i f(\bar{x}_j; \hat{\mu}_i, \hat{\sigma}^2/r_j) \bigg/ \left\{\sum_{t=1}^{g} \pi_t f(\bar{x}_j; \hat{\mu}_t, \hat{\sigma}^2/r_j)\right\} \tag{6.2.6}$$

is the estimated posterior probability that treatment $j$ belongs to $G_i$ $(i = 1, \ldots, g)$. It follows from the explanation in Section 1.6 on likelihood estimation for mixture models in general, that $\hat{\phi}$ can be computed iteratively, using some initial estimate of $\phi$ in the right-hand side of the above equations to produce a new estimate of $\phi$, and so on. A FORTRAN listing is supplied in the Appendix. For this problem there is an obvious initial estimate of $\sigma^2$ in $s^2$, while an initial estimate of $\mu_i$ can be computed on the basis of some partition of the treatment sample means arising from consideration of, say, the least significant differences following the analysis of variance. Various partitions should be considered for each specified $g$.

We now examine whether the solution of (6.2.3) to (6.2.5) provides the maximum likelihood estimate of $\phi$ for a normal mixture model fitted to all the data available in the case of a completely randomized design or a randomized complete block design (RCBD). For a completely randomized design, where each treatment $j$ is replicated $r_j$ times ($j = 1, \ldots, n$), the model is of the form $x_{jk}$, where $x_{jk}$ refers to the $k$th replicate on treatment $j$ ($j = 1, \ldots, n$; $k = 1, \ldots, r_j$). Thus

$$\bar{x}_j = \sum_{k=1}^{r_j} x_{jk}/r_j \qquad (j = 1, \ldots, n),$$

$$s^2 = \sum_{j=1}^{n} \sum_{k=1}^{r_j} (x_{jk} - \bar{x}_j)^2/\nu,$$

and $\nu = \sum_{j=1}^{j=n} r_j - n$. It is easy to verify that the maximum likelihood estimate of $\phi$ obtained by fitting a mixture of $g$ univariate normal distributions to this design can be expressed in the form (6.2.3) to (6.2.5).

For equireplicated treatments in the form of a randomized complete block design with fixed block effect $\varsigma_k$ corresponding to the $k$th block ($k = 1, \ldots, r$), $x_{jk}$ denotes the replicate of treatment $j$ in the $k$th block ($j = 1, \ldots, n$; $k = 1, \ldots, r$), and

$$\bar{x}_j = \sum_{k=1}^{r} x_{jk}/r \qquad (j = 1, \ldots, n),$$

and

$$\bar{x} = \sum_{j=1}^{n} \bar{x}_j/n.$$

The analysis of variance provides

$$s^2 = \sum_{j=1}^{n} \sum_{k=1}^{r} (x_{jk} - \bar{x}_j - \hat{\varsigma}_k)^2/\nu$$

as an estimate of $\sigma^2$, where $\nu = (n-1)(r-1)$ and

$$\hat{\varsigma}_k = \sum_{j=1}^{n} x_{jk}/n - \bar{x}, \tag{6.2.7}$$

which can be interpreted as the estimate of the fixed effect $\varsigma_k$ of the $k$th block under the constraint

$$\sum_{k=1}^{r} \varsigma_k = 0.$$

In fitting a normal mixture model to the data $x_{jk}$ in this design, the maximum likelihood estimates of $\pi$ and $\mu_1, \ldots, \mu_g$ satisfy (6.2.3) and (6.2.4). The maximum likelihood estimate of $\sigma^2$ is given by (6.2.5) if the divisor of $\nu + n$, which equals $nr - (r-1)$ for the randomized block design, is replaced by $nr$. For random block effects, however, the individual observations $x_{jk}$ are needed if a normal mixture model is to be fitted according to maximum likelihood. This is discussed in the next section.

## 6.3 FITTING OF A NORMAL MIXTURE MODEL TO A RCBD WITH RANDOM BLOCK EFFECTS

We consider here the fitting of a normal mixture model to data from a randomized complete block design, where the block effects $\varsigma_k$ are taken to be independent and identically distributed according to

$$\varsigma_k \sim N(0, \xi^2) \qquad (k = 1, \ldots, r).$$

Observations on treatments in the same block are no longer independent as there is a covariance $\xi^2$ between them. As a consequence, the computation of the maximum likelihood estimate of

$$\phi = (\pi', \mu_1, \ldots, \mu_g, \sigma^2, \xi^2)'$$

is rather cumbersome, as shown by Basford and McLachlan (1985b). However, one advantage of adopting this model over the more straightforward conditional approach with fixed block effects is that it appears to give less extreme estimates of the posterior probabilities of group membership.

Basford and McLachlan (1985b) showed that the maximum likelihood estimate of $\phi$ satisfies (6.2.3) and

$$\sum_{jk} \hat{\tau}_{ij}(x_{jk} - \hat{\mu}_i)$$

$$+ r\hat{\xi}^2 \left\{ -n\bar{x} \sum_{j} \hat{\tau}_{ij} + \sum_{i'jj'} \hat{\tau}_{iji'j'}\hat{\mu}_{i'} \right\} \Big/ (\hat{\sigma}^2 + n\hat{\xi}^2)$$

$$= 0 \qquad (i = 1, \ldots, g), \tag{6.3.1}$$

$$\hat{\sigma}^2 = \left[ \sum_{ijk} \hat{\tau}_{ij}(x_{jk} - \hat{\mu}_i)^2 - n \left\{ \sum_{k} \hat{\varsigma}_k^2 + ra \right\} \right] / r(n-1), \tag{6.3.2}$$

and

$$\hat{\xi}^2 = \sum_{k} \hat{\varsigma}_k^2 / r + a - \hat{\sigma}^2/n, \tag{6.3.3}$$

where

$$a = \bar{x}^2 - 2\bar{x} \sum_{ij} \hat{\tau}_{ij}\hat{\mu}_i/n + \sum_{iji'j'} \hat{\tau}_{iji'j'}\hat{\mu}_i\hat{\mu}_{i'}/n^2$$

and where the notation $\hat{\varsigma}_k$ in the fixed effects model is still used for (6.2.7). In these equations, $\hat{\tau}_{iji'j'}$ is the estimate of the joint posterior probability that treatments $j$ and $j'$ belong to $G_i$ and $G_{i'}$, respectively and, as before, $\hat{\tau}_{ij}$ is the estimated posterior probability that treatment $j$ belongs to $G_i$, although it is no longer given by (6.2.6). Here the solution of the M step of the EM algorithm does not exist in closed form, and the equations (6.3.1) to (6.3.3) must be solved iteratively. Note that if a negative value of (6.3.3) is obtained, then the maximum likelihood estimate of $\xi^2$ is zero.

The amount of calculation involved in computing the $\tau_{ij}$ and $\tau_{iji'j'}$ at each stage of the iterative process is considerable, even for a moderate number of treatments. To calculate $\tau_{ij}$, first condition on the block effects and then finally take the expectation with respect to these effects. This leads after some algebra to

$$\tau_{ij} = \frac{\sum\limits_{\substack{t_1 \ldots t_n \\ t_j = i}} \pi_{t_1} \cdots \pi_{t_n} \exp(v_{t_1 \ldots t_n})}{\sum\limits_{t_1 \ldots t_n} \pi_{t_1} \cdots \pi_{t_n} \exp(v_{t_1 \ldots t_n})}, \tag{6.3.4}$$

where

$$v_{t_1 \ldots t_n} = -\tfrac{1}{2} r\sigma^{-2} \left\{ \sum_{j=1}^{n} v_{t_j}^2 - \xi^2(\sigma^2 + n\xi^2)^{-1} \left( \sum_{j=1}^{n} v_{t_j} \right)^2 \right\},$$

and

$$v_{t_j} = \bar{x}_j - \mu_{t_j} \qquad (j = 1, \ldots, n).$$

Similarly, for $j \neq j'$ it can be shown that

$$\tau_{iji'j'} = \frac{\displaystyle\sum_{\substack{t_1 \ldots t_n \\ t_j = i, t_{j'} = i'}} \pi_{t_1} \ldots \pi_{t_n} \exp(v_{t_1 \ldots t_n})}{\displaystyle\sum_{t_1 \ldots t_n} \pi_{t_1} \ldots \pi_{t_n} \exp(v_{t_1 \ldots t_n})}. \tag{6.3.5}$$

For $j = j'$, $\tau_{iji'j'}$ is obviously equal to $\tau_{ij}$ for $i = i'$, and zero otherwise.

The posterior probabilities depend on all the treatment means, rather than just on the appropriate treatment mean as in equation (6.2.6). In equation (6.3.4), there are $n$ summations in the denominator and $n-1$ (since $t_j$ is set equal to $i$) summations in the numerator, each over the range 1 to $g$, while in equation (6.3.5), there are $n-2$ summations in the numerator over the same range. Thus $\tau_{ij}$ and $\tau_{iji'j'}$ are computationally feasible only for a relatively small number of treatments. The same problems occur with the calculation of the log likelihood under the mixture model, which can be expressed as

$$\begin{aligned} L(\phi) = & -\tfrac{1}{2} nr \log(2\pi) - \tfrac{1}{2} r(n-1) \log \sigma^2 - \tfrac{1}{2} r \log(\sigma^2 + n\xi^2) \\ & - \tfrac{1}{2}\sigma^{-2} \sum_{jk} (x_{jk} - \bar{x}_j)^2 + \tfrac{1}{2} n^2 \xi^2 \sigma^{-2} (\sigma^2 + n\xi^2)^{-1} \sum_k \hat{\varsigma}_k^2 \\ & + \log \left\{ \sum_{t_1 \ldots t_n} \pi_{t_1} \ldots \pi_{t_n} \exp(v_{t_1 \ldots t_n}) \right\}. \end{aligned}$$

Binder (1978, 1981), with his Bayesian approach to cluster analysis, was faced with a similar problem of it not being feasible to compute the posterior probability of each possible grouping. He used an approximation obtained by a searching algorithm which sampled partitions to identify those with high posterior probabilities, and subsequently worked with those. Basford and Horton (1984) considered a method for estimating the likelihood for the random effects model using a sampling procedure. It consisted of a searching algorithm to identify those partitions with high posterior probabilities, with intensive sampling confined to regions about those partitions. However, when it is computationally prohibitive to implement the mixture likelihood approach with random block effects, the simplest way to proceed is to fit the mixture model by conditioning on the blocks; that is, by proceeding as for a randomized complete block design with fixed block effects.

## 6.4 SOME OTHER METHODS OF PARTITIONING TREATMENT MEANS

Cox and Spjøtvoll (1982) proposed a simple method based solely on standard $F$ tests for assessing $g$ and the group membership of the $n$ treatments as represented by $\mathbf{Z} = (\mathbf{z}_1', \ldots, \mathbf{z}_n')'$. Their method is appropriate for evaluating consistency with the hypothesis $H_{\mathbf{Z}}$ corresponding to a specification of $\mathbf{Z}$ for any $g$ from 1 to $n$.

Consistency with the hypothesis $H_{\mathbf{Z}}$ is judged on the basis of the observed value of the statistic

$$W(g)/\left\{(n-g)s^2\right\}, \tag{6.4.1}$$

where

$$W(g) = \sum_{i=1}^{g}\sum_{j=1}^{n} z_{ij} r_j (\bar{x}_j - \tilde{\mu}_i)^2$$

is the sum of squares within groups and

$$\tilde{\mu}_i = \sum_{j=1}^{n} z_{ij} r_j \bar{x}_j \Bigg/ \left(\sum_{j=1}^{n} z_{ij} r_j\right)$$

is the weighted average of the treatment sample means in $G_i$. The sum of squares within groups depends on $g$ through $\mathbf{Z}$ which is suppressed in our notation $W(g)$. The $P$-value for the hypothesis $H_{\mathbf{Z}}$ is given by the area to the right of (6.4.1) under the $F_{n-g,\nu}$ distribution. For a conventional level of significance, say $\alpha = 0.01$, 0.05, or 0.1, hypotheses which are consistent with the data are considered. They correspond to partitions $\mathbf{Z}$ with a $P$-value greater than the proposed level of significance. The method provides a conservative lower $1-\alpha$ confidence limit for the "true" number of groups.

The first hypothesis tested is $g = 1$, for which (6.4.1) is the usual $F$ statistic for testing the homogeneity of all the $n$ treatment means; $W(1)$ is the sum of squares between treatments in the analysis of variance. If this hypothesis is rejected, the case $g = 2$ is considered for which there are $2^{n-1} - 1$ distinct possible partitions of the means into two groups given by all admissible values of $\mathbf{Z}$. The number of groups $g$ is increased until a partition is found to be consistent with the data; that is, $P > 0.05$ or 0.1.

Cox and Spjøtvoll (1982) suggested the following approach in applying their method.

(i) Obtain the smallest number $g$ such that the data are consistent with $g$ groups of different means. Then at least $g$ groups are needed to satisfactorily describe the data.
(ii) List the partitions into $g$ groups that are consistent with the data.
(iii) Consider whether partitions into $g+1$ or more groups are appropriate to provide "refinements" of those partitions into $g$ groups.

For (ii) where there may be several partitions into $g$ groups consistent with the data, a conclusion which can be reported in a simple form concerns those sets of treatments for which members of a given set always fall in the same group and members of different sets always fall in different groups.

Concerning a clustering approach to this problem of partitioning treatment means, Scott and Knott (1974) used the hierarchical, divisive clustering method of Edwards and Cavalli-Sforza (1965) based on $W(g)$, or equivalently, the sum of squares between groups. In proceeding initially from $g = 1$ to $g = 2$, the partition $\mathbf{Z}$ is sought for which $W(2)$ is a minimum. An approximation to the asymptotic distribution of

$$\min_{\mathbf{Z}} W(2)/s^2$$

under $g = 1$ was given for computing an approximate $P$-value corresponding to the optimal partition. Subsequently partitions into three or more groups are obtained by applying this splitting procedure separately to each group of the partition for $g = 2$, and so on.

Caliński and Corsten (1985) proposed two clustering methods for the partitioning of treatment means in an analysis of variance in the case of equireplicated treatments ($r_j = r$ for $j = 1, \ldots, n$). One of the methods was embedded in an extension of a Studentized range simultaneous test procedure. With this test, the overall test of homogeneity, that is $H_0 : g = 1$, involves the comparison of the range $R_0 = |\bar{x}_{(n)} - \bar{x}_{(1)}|$ with

$$R_\alpha = (s^2/r)^{1/2} q_{n,\nu;1-\alpha}, \qquad (6.4.2)$$

where $\bar{x}_{(1)}, \ldots, \bar{x}_{(n)}$ denote the sample means in increasing order and $q_{n,\nu;1-\alpha}$ denotes the quantile of order $1-\alpha$ of the distribution of the Studentized range $q_{n,\nu}$ for $n$ sample means and $\nu$ degrees of freedom. The clustering method proposed in connection with this test is a hierarchical agglomerative procedure with ordinary distance used to compute the dissimilarity between any two means and complete linkage (furthest neighbor) method used to measure the distance between clusters. At the start the two sample means at closest distance are fused. At each following step, those two clusters whose range is smallest among all pairs of contiguous clusters are fused. At each step $i$, the smallest range $R_i$ is compared with $R_\alpha$ and, if it exceeds this threshold, then this procedure ends at the previous fusion.

By implementing this procedure for a fixed $\alpha$, there is the protection that the probability of accepting a value of $g$ greater than its "true" value is at most $\alpha$, and that if the overall null hypothesis $H_0$ is true, then this probability is precisely $\alpha$. This also applies to their second method to be described shortly. However, instead of keeping $\alpha$ fixed, Caliński and Corsten (1985) suggest monitoring the $P$-value at each step $i$, given by

$$\mathrm{pr}\left\{q_{n,\nu} \geq R_i(s^2/r)^{-1/2}\right\}.$$

It is a nonincreasing function of the number of fusions and there is usually a sharp drop in its value at the step before which the procedure would be terminated according to (6.4.2). As a consequence, Caliński and Corsten (1985) prefer the clustering obtained before the final one.

Their second clustering method is embedded in the extended $F$ ratio simultaneous test procedure of Gabriel (1964). For a given number of groups $g$, the partition $\mathbf{Z}$ is sought for which the sum of squares within groups $W(g)$ is a minimum. This value of $W(g)$ is nonincreasing with $g$ and the procedure can be terminated as soon as a partition is found for which $W(g)$ is less than

$$(n-1)s^2 F_{n-1,\nu;1-\alpha}. \tag{6.4.3}$$

Following a significant result for the usual $F$ test for $g = 1$, the procedure is implemented starting with $g = 2$.

As with their other procedure, Caliński and Corsten (1985) recommended the monitoring of the $P$-value at each step $g$, defined by

$$pr\left\{\mathcal{F}_{n-1,\nu} \geq \min_{\mathbf{Z}} W(g)/(n-1)s^2\right\} \tag{6.4.4}$$

where $\mathcal{F}_{n-1,\nu}$ denotes a $F_{n-1,\nu}$ distributed random variable. This $P$-value is a nondecreasing function of $g$ and a dramatic increase will usually be observed at some $g$ at which the procedure can be terminated. Caliński and Corsten (1985) suggested this as an alternative to termination based on (6.4.3). Note that they have proposed the use of $W(g)$ here as the basis of a simultaneous test of all hypotheses concerning the equality of two or more of the treatment means so that the term $(n-1)$ in (6.4.4) is fixed as the value of $g$ is increased from $g = 1$. By contrast, with the use of $W(g)$ in the partitioning method proposed by Cox and Spjøtvoll (1982), $(n-1)$ is updated to $(n-g)$ as in (6.4.1) to reflect the current degrees of freedom for $W(g)$ in the computation of the associated $P$-value for a given partition $\mathbf{Z}$ at each level of $g$ considered.

## 6.5 EXAMPLE 1

Some numerical examples are presented in this and the remaining sections of the chapter to demonstrate the usefulness of the mixture likelihood approach to the partitioning of treatment means. The first three of the four examples concern real data sets for which analyses by other methods have been reported in the literature on this problem. In the previous sections the treatments have been labeled 1 to $n$ with sample means $\bar{x}_1$ to $\bar{x}_n$. In the following examples they are labeled $a$, $b$, $c$, ... in order of increasing sample mean.

The first example considers a data set on fat absorption in doughnut mixtures as analysed by Cox and Spjøtvoll (1982) and, more recently, by Basford and McLachlan (1985b). Eight different doughnut mixtures were used on each of six consecutive days. The 48 individual observations are tabulated in Anderson and Bancroft (1952, page 245); the sample means are given below:

| $a$ | $b$ | $c$ | $d$ | $e$ | $f$ | $g$ | $h$ |
|---|---|---|---|---|---|---|---|
| 161 | 162 | 165 | 172 | 175 | 178 | 181 | 185 |

An analysis of variance gave $s^2 = 141.6$ with $\nu = 40$ degrees of freedom as an estimate of $\sigma^2$ and a $P$-value of 0.004 for the $F$ test of homogeneity of all the treatment means.

Basford and McLachlan (1985b) clustered the treatments by fitting a normal mixture model to this randomized complete block design. As the individual observations were known and there were only $n = 8$ treatments, it was possible to consider random as well as fixed block effects. In each case the increase in the log likelihood $L(\hat{\phi})$ as $g$ increases from one indicates that $g = 2$ is the smallest value compatible with the data; see Table 24. If the approximation of Wolfe (1971) is used for the null distribution of $-2\log\lambda$, that is $\chi_2^2$ here, then for the test of $g = 2$ against $g = 3$, the $P$-value is equal to 0.30 and 0.18 for random and fixed block effects respectively. As cautioned in Section 1.10, the $P$-value based on this approximation should not be rigidly interpreted, but used only as a guide to the smallest value of $g$ consistent with the data.

The null hypothesis of $g = 2$ versus the alternative of $g = 3$ is examined further now by bootstrapping the log likelihood ratio statistic to provide a test of a nominal size. This test, where $-2\log\lambda$ for the original data is compared with the value of the appropriate order statistic of $K$ subsequent bootstrap replications of it, has been described in Section 1.10 and already applied in the case of unreplicated data in some examples in Chapter 3. In the case of fixed block effects, the value of 3.43 for $-2\log\lambda$ formed from the original data is well short of the largest value of 8.34 for the $K = 19$ bootstrap replications generated for this statistic. Formally, the null hypothesis of $g = 2$ groups is not rejected at the 0.05 level. It would be retained too at the 0.1 level, as the original value of $-2\log\lambda$ is also less than one other bootstrap replication equal to 5.84. This last conclusion was reached also for this test applied in the case of random block effects, again using $K = 19$ bootstrap replications of $-2\log\lambda$.

Estimates of the posterior probabilities of group membership for $g = 2$ under both random and fixed block effects are displayed in Table 25. Both give the same grouping of $abc$ and $defgh$, when each treatment is assigned to the group in which its estimated posterior probability is a maximum. However, the estimated posterior probabilities for treatment $d$, obtained by treating the block effects as random, indicate that $d$ has no strong link with either of the two possible groups; $a$, $b$, and $c$ are clearly put in one group and $e$, $f$, $g$, and $h$ in the other. It has been observed also with

**TABLE 24** Example 1: Log likelihood under a normal mixture model for a RCBD with

(i) random block effects
(ii) fixed block effects

| $g$ | $L(\hat{\phi})$ (i) | $L(\hat{\phi})$ (ii) |
|---|---|---|
| 1 | −188.62 | −180.35 |
| 2 | −178.07 | −167.70 |
| 3 | −176.87 | −165.98 |

other data sets that the estimates of the posterior probabilities tend to be less extreme when the block effects are treated as random rather than fixed. This is not surprising, as treating the block effects as random can be viewed as adopting a Bayesian approach, at least for the block location parameter, and it has been noted with respect to estimation of posterior probabilities in a discriminant analysis context, that a Bayesian approach generally leads to less extreme estimates than those formed by a frequentist approach; see, for example, Aitchison, Habbema and Kay (1977), Desu and Geisser (1973), McLachlan (1979), and Rigby (1982).

The acceptance of $g = 2$ groups with the mixture likelihood approach agrees with the conclusion of Cox and Spjøtvoll (1982) that two groups are

**TABLE 25** Example 1: Estimated posterior probabilities $\hat{\tau}_{ij}$ of group membership for $g = 2$ for a RCBD with

(i) random block effects
(ii) fixed block effects

| | Treatment $j$ | $\hat{\tau}_{1j}$ | $\hat{\tau}_{2j}$ |
|---|---|---|---|
| (i) | $a$ | 1.00 | 0.00 |
| | $b$ | 1.00 | 0.00 |
| | $c$ | 1.00 | 0.00 |
| | $d$ | 0.33 | 0.67 |
| | $e$ | 0.00 | 1.00 |
| | $f$ | 0.00 | 1.00 |
| | $g$ | 0.00 | 1.00 |
| | $h$ | 0.00 | 1.00 |
| (ii) | $a$ | 1.00 | 0.00 |
| | $b$ | 1.00 | 0.00 |
| | $c$ | 1.00 | 0.00 |
| | $d$ | 0.10 | 0.90 |
| | $e$ | 0.00 | 1.00 |
| | $f$ | 0.00 | 1.00 |
| | $g$ | 0.00 | 1.00 |
| | $h$ | 0.00 | 1.00 |

Reprinted from K.E. Basford and G.J. McLachlan (1985b), *Commun. Statist.-Theor. Meth.* **14**, 451-463. By courtesy of Marcel Dekker, Inc.

adequate. They found that for a required probability $P$ greater than 0.1 for acceptable consistency with the data, the possible partitions into two groups were (the value of $P$ is in parentheses):

| | | |
|---|---|---|
| $ab$, | $cdefgh$ | (0.12) |
| $abc$, | $defgh$ | (0.58) |
| $abcd$, | $efgh$ | (0.53) |
| $abcde$, | $fgh$ | (0.25) |
| $abcdef$, | $gh$ | (0.10) |
| $abce$, | $dfgh$ | (0.15) |
| $abcdf$, | $egh$ | (0.13) |

It can be seen that the partition with the highest value of $P$ agrees with that obtained by the mixture likelihood approach. Concerning the other possible partitions, only the partition with the second highest value of $P$, which has treatment $d$ grouped with $a$, $b$, and $c$ rather than with $e$, $f$, $g$, and $h$, is supported to any extent by the estimated posterior probabilities of group membership.

## 6.6 EXAMPLE 2

The barley data discussed by Duncan (1955) have often been used in the literature to illustrate techniques of grouping treatments. There are seven varieties of barley grown in $r = 6$ blocks. The sample means in bushels per acre are listed below:

| $a$ | $b$ | $c$ | $d$ | $e$ | $f$ | $g$ |
|---|---|---|---|---|---|---|
| 49.6 | 58.1 | 61.0 | 61.5 | 67.6 | 71.2 | 71.3 |

An analysis of variance gave $s^2 = 79.64$ with $\nu = 30$ degrees of freedom and a $P$-value of 0.002 for the $F$ test of homogeneity of all the treatment means.

A mixture of $g = 2$ normal densities

$$f(x; \phi) = \pi_1 f(x; \mu_1, \sigma^2/r) + \pi_2 f(x; \mu_2, \sigma^2/r), \tag{6.6.1}$$

was fitted to the seven sample means $\bar{x}_a$ to $\bar{x}_g$ from this randomized complete block design, assuming fixed block effects. A model with random block effects was unable to be fitted as well, as information on the individual observations of this experiment was not available to the authors. The maximum likelihood estimate of

$$\phi = (\pi_1, \pi_2, \mu_1, \mu_2, \sigma^2)'$$

computed on the basis of $\bar{x}_a$ to $\bar{x}_g$ and also $s^2$ is

$$\hat{\phi} = (0.54, 0.46, 57.40, 69.35, 83.77)'.$$

A mixture of $g = 3$ normal densities was fitted also. For a given $g$ ($g = 1$ to 3), the value of the log likelihood $L(\hat{\phi})$ is listed in Table 26, along with the $P$-value for the likelihood ratio test of $g$ versus $g+1$ groups, computed according to the $\chi_2^2$ approximation to the null distribution of $-2\log\lambda$. The estimated posterior probabilities of group membership are displayed in Table 27 for $g = 2$ and 3. They clearly suggest the partition *abcd* and *efg* for $g = 2$ and *a*, *bcd*, and *efg* for $g = 3$.

On the choice of $g$, the $P$-values in Table 26 indicate $g = 2$ but, as stressed earlier, they are approximate values and serve only as a guide. The null hypothesis of $g = 2$ versus the alternative of $g = 3$ is considered further now, using $K = 19$ bootstrap replications of $-2\log\lambda$ to carry out a test with a significance level of 0.05 in mind. With this example each bootstrap sample consists of observations on $\bar{x}_a^*$ to $\bar{x}_g^*$ and $s^{*2}$. Before being ordered, $\bar{x}_a^*$ to $\bar{x}_g^*$ are distributed independently according to the null density (6.6.1) for the treatment means but with $\hat{\phi}$ in place of $\phi$, each independent of $s^{*2}$ distributed as

$$\nu s^{*2}/\hat{\sigma}^2 \sim \chi_\nu^2.$$

**TABLE 26** Example 2: Log likelihood and $P$-value for likelihood ratio test of $g$ versus $g+1$ groups under a normal mixture model

| $g$ | $L(\hat{\phi})$ | $P$ |
|---|---|---|
| 1 | −71.76 | 0.034 |
| 2 | −68.37 | 0.301 |
| 3 | −67.17 | 1.000 |

**TABLE 27** Example 2: Estimated posterior probabilities $\hat{\tau}_{ij}$ of group membership for $g = 3$ and in parentheses for $g = 2$

| Treatment $j$ | $\hat{\tau}_{1j}$ | $\hat{\tau}_{2j}$ | $\hat{\tau}_{3j}$ |
|---|---|---|---|
| $a$ | 0.98<br>(1.00) | 0.02<br>(0.00) | 0.00 |
| $b$ | 0.02<br>(0.99) | 0.98<br>(0.01) | 0.00 |
| $c$ | 0.00<br>(0.90) | 0.97<br>(0.10) | 0.03 |
| $d$ | 0.00<br>(0.85) | 0.95<br>(0.15) | 0.05 |
| $e$ | 0.00<br>(0.03) | 0.13<br>(0.97) | 0.87 |
| $f$ | 0.00<br>(0.00) | 0.01<br>(1.00) | 0.99 |
| $g$ | 0.00<br>(0.00) | 0.01<br>(1.00) | 0.99 |

A mixture of $g$ normal densities was fitted to each bootstrap sample for $g = 2$ and 3 in turn, and twice the increase in the log likelihood for $g = 3$ over $g = 2$ was calculated. As the value of 2.40 for $-2\log\lambda$ formed from the original data is less than two of the bootstrap values of 2.53 and 2.81 for this statistic, the null hypothesis of $g = 2$ groups is not rejected at the 0.1 level.

Concerning other analyses of this data set, the hierarchical splitting method of Scott and Knott (1974) gave the same partitions of *abcd* and *efg*, and *a*, *bcd*, and *efg* for $g = 2$ and 3 respectively, as obtained above under a normal mixture model. On the question of the number of groups, the split from $g = 2$ to $g = 3$ groups was found to be on the borderline of significance. The Bayesian approach of Binder (1978, 1981) isolated the solution to only three possibilities, one group, the $g = 2$ group partition *abcd* and *efg*, and the $g = 3$ group partition *a*, *bcd*, and *efg*. These last two partitions were suggested by the frequentist approach to the fitting of a normal mixture model. As reported by Binder (1981), "which partition is most suitable depends on our *a priori* assessment of how far apart the groups should be." Unless the prior for $g$ strongly favours one group, the

**TABLE 28** Example 2: Results for agglomerative method of Caliński and Corsten

| Fusion | *f-g* | *c-d* | *b-d* | *e-g* | *a-d* | *a-g* |
|---|---|---|---|---|---|---|
| *P* | 1.000 | 1.000 | 0.994 | 0.990 | 0.272 | 0.004 |

Reproduced from T. Caliński and L.C.A. Corsten (1985), Clustering means in ANOVA by simultaneous testing. *Biometrics* **41**, 39-48. With permission from the Biometric Society.

other two partitions into two or three groups were usually obtained. The partition into two groups generally resulted if the prior assumptions implied a wide dispersion of the group means relative to $\sigma^2$.

The same partitions as above for $g = 2$ and 3 groups were obtained also by Caliński and Corsten (1985) with each of their two clustering procedures. In Table 28, we report their results for the fusions and associated $P$-values for their agglomerative method based on the Studentized range. For their other method based on $W(g)$, we give in Table 29 the partition which minimizes $W(g)$ and the associated $P$-values for $g = 1$ and 2. Caliński and Corsten (1985) noted that their agglomerative method is terminated at the $g = 2$ step, using the critical range (6.4.2) with $\alpha = 0.05$. However, it can be seen from Table 28, that if we use their other criterion for termination, namely the value of $g$ at which there is a sharp drop in the $P$-value, the procedure is terminated at the previous step of $g = 3$. Their other method points to $g = 2$, no matter whether the procedure is terminated either according to (6.4.3) for $\alpha = 0.05$ (that is, $W(g) < 1156.37$) or at the value of $g$ at which there is an appreciable increase in the $P$-value.

**TABLE 29** Example 2: Results for method of Caliński and Corsten based on $W(g)$

| $g$ | Partition | Minimum of $W(g)$ | $P$ |
|---|---|---|---|
| 1 | *a* to *g* | 2202.24 | 0.002 |
| 2 | *abcd*, *efg* | 599.38 | 0.308 |

Reproduced from T. Caliński and L.C.A. Corsten (1985), Clustering means in ANOVA by simultaneous testing. *Biometrics* **41**, 39-48. With permission from the Biometric Society.

Cox and Spjøtvoll (1982) also analyzed this barley data set. For $g = 2$, they found there were five possible partitions with $P > 0.05$, as listed below with the $P$-value in parentheses.

| | | |
|---|---|---|
| $a$, | $bcdefg$ | (0.059) |
| $ab$, | $cdefg$ | (0.088) |
| $abc$, | $defg$ | (0.11) |
| $abcd$, | $efg$ | (0.22) |
| $abd$, | $cefg$ | (0.082) |

The partition with the highest $P$-value corresponds to the $g = 2$ partition produced by the other methods discussed above. They also listed some partitions for $g = 3$, "partly because none of the partitions into two groups gives an excellent fit." The partition into three groups with the highest $P$-value ($P = 0.88$) is $a$, $bcd$, and $efg$, which is the same as the $g = 3$ partition produced by the other methods considered here.

## 6.7 EXAMPLE 3

In this example we use the data analyzed by Keuls (1952) in his proposal for a Studentized range test and recently reanalyzed by Caliński and Corsten (1985). The data concern a trial on the mean gross weight in grams per head of thirteen white cabbage varieties in three complete blocks. The sample means in coded form are

| $a$ | $b$ | $c$ | $d$ | $e$ | $f$ | $g$ |
|---|---|---|---|---|---|---|
| 97.7 | 100.7 | 111.3 | 120.7 | 124.3 | 128.7 | 129.0 |

| $h$ | $i$ | $j$ | $k$ | $l$ | $m$ |
|---|---|---|---|---|---|
| 131.0 | 132.0 | 141.7 | 150.7 | 152.7 | 176.0 |

An analysis of variance gave $s^2 = 124.29$ with $\nu = 24$ degrees of freedom as an estimate of $\sigma^2$ and a practically zero $P$-value for the $F$ test of homogeneity of all the treatment means.

For various levels of $g$, a normal mixture model was fitted to these 13 sample means $\bar{x}_a$ to $\bar{x}_m$ from this randomized complete block design, assuming fixed block effects. In Table 30 we report for a given $g$ ($g = 1$ to 4), the value of the log likelihood $L(\hat{\phi})$ formed from $\bar{x}_a$ to $\bar{x}_m$ and $s^2$, along with the $P$-value for the likelihood ratio test of $g$ versus $g+1$ groups, computed using the $\chi_2^2$ approximation to the null distribution of $-2\log\lambda$. The values suggest that at least $g = 3$, and more likely $g = 4$, groups are

**TABLE 30**
Example 3: Log likelihood and *P*-value for likelihood ratio test of $g$ versus $g+1$ groups under a normal mixture model

| $g$ | $L(\hat{\phi})$ | $P$ |
|---|---|---|
| 1 | −103.89 | 0.001 |
| 2 | −97.09 | 0.008 |
| 3 | −92.28 | 0.036 |
| 4 | −88.94 | 0.991 |

needed. Support for the latter number of groups is provided also by a test of $g = 3$ versus $g = 4$, using $K = 19$ bootstrap replications of $-2\log\lambda$. Carrying out this test as in the previous example, the value of 6.68 for $-2\log\lambda$ formed from the original sample well exceeds the largest value (2.77) of its 19 bootstrap replications and, so formally, $g = 3$ is rejected at the 0.05 level.

The estimated posterior probabilities of group membership are displayed in Table 31 for $g = 3$ and 4. For $g = 3$, they clearly suggest the partition

$$abc,\ defghij,\ klm.$$

The partition into four groups on the basis of the estimated posterior probabilities is

$$abc,\ defghi,\ jkl, m,$$

although there is some doubt as to whether treatment $j$ belongs to the second or third groups.

Concerning the two methods proposed by Caliński and Corsten (1985), they both lead to the same grouping for $g = 4$ as the above outright assignment of the treatments to four groups on the basis of the estimated posterior probabilities under the mixture likelihood approach. Also, for $g = 3$, their method based on the sum of squares within groups $W(g)$ gave the same partition of the treatments as with the mixture approach. However, their

**TABLE 31** Example 3: Estimated posterior probabilities $\hat{\tau}_{is}$ of group membership for $g = 4$ and in parentheses for $g = 3$

| Treatment $s$ | $\hat{\tau}_{1s}$ | $\hat{\tau}_{2s}$ | $\hat{\tau}_{3s}$ | $\hat{\tau}_{4s}$ |
|---|---|---|---|---|
| $a$ | 1.00 | 0.00 | 0.00 | 0.00 |
| | (1.00) | (0.00) | (0.00) | |
| $b$ | 1.00 | 0.00 | 0.00 | 0.00 |
| | (1.00) | (0.00) | (0.00) | |
| $c$ | 0.90 | 0.10 | 0.00 | 0.00 |
| | (0.86) | (0.14) | (0.00) | |
| $d$ | 0.01 | 0.99 | 0.00 | 0.00 |
| | (0.03) | (0.97) | (0.00) | |
| $e$ | 0.00 | 1.00 | 0.00 | 0.00 |
| | (0.00) | (1.00) | (0.00) | |
| $f$ | 0.00 | 1.00 | 0.00 | 0.00 |
| | (0.00) | (1.00) | (0.00) | |
| $g$ | 0.00 | 1.00 | 0.00 | 0.00 |
| | (0.00) | (1.00) | (0.00) | |
| $h$ | 0.00 | 1.00 | 0.00 | 0.00 |
| | (0.00) | (1.00) | (0.00) | |
| $i$ | 0.00 | 0.99 | 0.01 | 0.00 |
| | (0.00) | (1.00) | (0.00) | |
| $j$ | 0.00 | 0.22 | 0.78 | 0.00 |
| | (0.00) | (0.93) | (0.07) | |
| $k$ | 0.00 | 0.00 | 1.00 | 0.00 |
| | (0.00) | (0.04) | (0.96) | |
| $l$ | 0.00 | 0.00 | 1.00 | 0.00 |
| | (0.00) | (0.01) | (0.99) | |
| $m$ | 0.00 | 0.00 | 0.00 | 1.00 |
| | (0.00) | (0.00) | (1.00) | |

agglomerative method based on the Studentized range test gave

$$abc,\ defghijkl,\ m$$

where treatments $k$ and $l$ are put with the middle rather than the third group. On the question of $g = 3$ versus $g = 4$ groups their agglomerative method would be terminated at $g = 3$ using (6.4.2) with $\alpha = 0.05$, but would be terminated at the previous step of $g = 4$ if the sharp drop in the $P$-value is adopted as the criterion for the selection of $g$. Their other method based on $W(g)$ would terminate at $g = 3$ for $W(g)$ compared to (6.4.3), but would proceed to $g = 4$ if the procedure is continued until a substantial increase in the $P$-value is observed.

## 6.8 EXAMPLE 4

In this final example of the chapter a set of simulated data on some treatment means is considered. One informative aspect of analysing such a set is that there is a true grouping of the treatments and so the effectiveness of any clustering can be observed. Data in the form of a randomized complete block design were generated from a mixture of three normal distributions which incorporated random block effects. Parameter values were taken equal to their sample analogues for the barley data set analysed in Section 6.6. Hence the block and error variances used in the simulation were 8.90 and 79.64 respectively, as obtained from the analysis of variance of this set. The means of the three different groups for the treatments were taken to be 49.6, 60.2, and 70.03, corresponding to the weighted average of the treatment sample means for each of the three groups produced by the mixture likelihood approach applied to this set of seven varieties of barley. There were seven treatments in each of six blocks (see Table 32), with treatments $a$ and $b$ from group $G_1$, $c$ and $d$ from $G_2$, and $e$, $f$, and $g$ from $G_3$.

As there are only $n = 7$ treatments, a normal mixture model is able to be fitted to this randomized complete block design with the block effects taken to be random. Here the number of underlying groups is known to

**TABLE 32** Data for Example 4: seven treatments in six blocks

| Treatment $j$ | $x_{j1}$ | $x_{j2}$ | $x_{j3}$ | $x_{j4}$ | $x_{j5}$ | $x_{j6}$ | $\bar{x}_j$ |
|---|---|---|---|---|---|---|---|
| $a$ | 41.40 | 49.96 | 48.35 | 65.85 | 62.36 | 44.13 | 52.01 |
| $b$ | 44.05 | 46.68 | 46.64 | 62.94 | 43.10 | 70.46 | 52.31 |
| $c$ | 44.72 | 42.92 | 60.54 | 72.92 | 59.10 | 60.53 | 56.79 |
| $d$ | 66.90 | 56.60 | 50.94 | 69.49 | 52.60 | 70.72 | 61.21 |
| $e$ | 64.33 | 60.94 | 69.59 | 68.90 | 56.98 | 69.65 | 65.07 |
| $f$ | 59.75 | 65.19 | 73.79 | 80.61 | 52.04 | 87.87 | 69.88 |
| $g$ | 69.94 | 52.10 | 76.57 | 78.05 | 60.29 | 83.24 | 70.03 |

be three, but it is of interest to see what conclusion about the number of groups is reached on the basis of the likelihood ratio test. The values of the log likelihood $L(\hat{\phi})$ for $g = 1$, 2, and 3 under both random and fixed block effects are shown in Table 33. For both models the likelihood ratio clearly points to $g = 2$ groups. Although it is rather disappointing that the likelihood ratio test fails to detect the presence of three groups, it can be explained by the fact that the groups are not widely separated from one another.

Estimates of the posterior probabilities of group membership for $g = 2$ for both random and fixed block effects are displayed in Table 34. Both give the same grouping of *abc* and *defg* where, however, the posterior probabilities of the treatments obtained by treating the block effects as random show that treatment *d* has no strong link with either of the two possible groups; treatments *a*, *b* and *c* are clearly put in one group and *e*, *f* and *g* in the other. As suggested in Section 1.10, it is often worthwhile examining the posterior probabilities for values of *g* other than that indicated by the likelihood ratio test. In Table 35, the posterior probabilities are displayed in the case of $g = 3$ groups for both random and fixed block effects. The results for both models suggest that *d* is distinct from the other treatments, and that it would be reasonable to have an additional group over $g = 2$ to which *d* can be assigned. By examining the estimated posterior probabilities in Tables 34 and 35, it is evident that these estimates are less extreme when the block effects are treated as random rather than fixed, as was the case in Example 1.

With $g = 3$, it can be seen from Table 35 that in terms of the estimated posterior probabilities of group membership, six out of the seven treatments

**TABLE 33** Example 4: Log likelihood under a normal mixture model for a RCBD with

(i) random block effects
(ii) fixed block effects

| $g$ | $L(\hat{\phi})$ | |
|---|---|---|
| | (i) | (ii) |
| 1 | −161.70 | −155.05 |
| 2 | −155.88 | −147.85 |
| 3 | −155.67 | −147.42 |

**TABLE 34** Example 4: Estimated posterior probabilities $\hat{\tau}_{ij}$ of group membership for $g = 2$ for a RCBD with

(i) random block effects
(ii) fixed block effects

| | Treatment $j$ | $\hat{\tau}_{1j}$ | $\hat{\tau}_{2j}$ |
|---|---|---|---|
| (i) | $a$ | 1.00 | 0.00 |
| | $b$ | 1.00 | 0.00 |
| | $c$ | 0.98 | 0.02 |
| | $d$ | 0.44 | 0.56 |
| | $e$ | 0.02 | 0.98 |
| | $f$ | 0.00 | 1.00 |
| | $g$ | 0.00 | 1.00 |
| (ii) | $a$ | 1.00 | 0.00 |
| | $b$ | 1.00 | 0.00 |
| | $c$ | 1.00 | 0.00 |
| | $d$ | 0.31 | 0.69 |
| | $e$ | 0.00 | 1.00 |
| | $f$ | 0.00 | 1.00 |
| | $g$ | 0.00 | 1.00 |

would be allocated correctly to the groups from which they were generated. Treatment $c$ would be incorrectly assigned to the first group, rather than to its true group of origin containing also $d$, although with random block effects, which is the appropriate model here, its estimated posterior probability of belonging to the second group was greater than that obtained by conditioning on block effects (0.27 versus 0.19). The generated sample mean of 56.79 for treatment $c$ is almost equidistant between the generated means of $b$ and $d$. Therefore, it is not surprising that $c$ is misallocated, although at a first glance one might anticipate that its estimated posterior probability of belonging to its true group of origin (the second group) would be closer to 0.5. Indeed, after the first iteration of the EM algorithm applied for random block effects the estimated posterior probability of its belonging to the second group was greater than 0.5 (0.62). However, as the iterations proceeded, the estimated mean of the first group increased towards the generated sample mean of treatment $c$, to a final value of 53.61, while the estimated mean of the second group increased away to a final value of 61.34; hence the final value of only 0.27 for the estimated posterior

**TABLE 35** Example 4: Estimated posterior probabilities $\hat{\tau}_{ij}$ of group membership for $g = 3$ for a RCBD with

(i) random block effects
(ii) fixed block effects

| | Treatment $j$ | $\hat{\tau}_{1j}$ | $\hat{\tau}_{2j}$ | $\hat{\tau}_{3j}$ |
|---|---|---|---|---|
| (i) | $a$ | 0.99 | 0.01 | 0.00 |
| | $b$ | 0.98 | 0.02 | 0.00 |
| | $c$ | 0.77 | 0.27 | 0.00 |
| | $d$ | 0.13 | 0.72 | 0.15 |
| | $e$ | 0.01 | 0.34 | 0.65 |
| | $f$ | 0.00 | 0.02 | 0.98 |
| | $g$ | 0.00 | 0.02 | 0.98 |
| (ii) | $a$ | 1.00 | 0.00 | 0.00 |
| | $b$ | 1.00 | 0.00 | 0.00 |
| | $c$ | 0.81 | 0.19 | 0.00 |
| | $d$ | 0.06 | 0.87 | 0.07 |
| | $e$ | 0.00 | 0.33 | 0.67 |
| | $f$ | 0.00 | 0.01 | 0.99 |
| | $g$ | 0.00 | 0.01 | 0.99 |

probability that $c$ belongs to the second group. These results correspond to starting the EM algorithm with initial values of the parameters computed with each treatment assigned to its true group of origin, so that $c$ was actually put with $d$ in the second group. The EM algorithm was implemented for other possible groupings of the treatments, but the same final estimates were obtained always.

Applying Cox and Spjøtvoll's (1982) method of partitioning means into groups gave the following groupings with their respective $P$-values in parentheses:

| | | | |
|---|---|---|---|
| *abc*, | | *defg* | (0.28) |
| *abcd*, | | *efg* | (0.25) |
| *ab*, | *cd*, | *efg* | (0.65) |
| *ab*, | *cde*, | *fg* | (0.51) |
| *abc*, | *d*, | *efg* | (0.57) |
| *abc*, | *de*, | *fg* | (0.71) |

These results agree with those of the mixture approach, in that they indicate two groups are adequate with *abc* in one and *efg* in the other, with some uncertainty as to which group *d* should belong. Considering the partitions for $g = 3$, the *P*-values indicate that *ab* is in one group, *d* is in another, and *fg* is in a third group. There is some doubt whether *c* should be in group one or two and also whether *e* should be in group two or three. The grouping produced by fitting a normal mixture model for $g = 3$ has the third highest *P*-value of the possible groupings at this level, while the actual grouping has the second highest *P*-value.

# 7
# Mixture Likelihood Approach to the Clustering of Three-Way Data

## 7.1 INTRODUCTION

In most approaches to clustering in the biological sciences, the basic data are viewed as a two-mode two-way array, where one of the modes is to be partitioned into groups on the basis of the other mode. To illustrate this, consider the data collected in a large plant improvement program, where an overall summary of the patterns of genotype response is often more useful than the traditional comparison of individual responses. In such experiments, two different types of two-mode two-way arrays are usually generated. The genotypes can be characterized by attributes producing a genotype by attribute ($G \times A$) matrix. They can also be characterized by the performance values for a single attribute measured in a number of environments, ($G \times E$) matrix. Many methods of clustering have been applied to such two-way arrays to provide a grouping of the genotypes; see, for example, Burt, Edye, Williams, Grof and Nicholson (1971), Mungomery, Shorter and Byth (1974), Byth, Eisemann and DeLacy (1976). Such analyses have been very useful to plant scientists. However, the restriction of being able to handle only two-way arrays has been a limitation; for example, where one is interested in the performance of genotypes with respect to several attributes in a number of environments.

Ideally, one would like to perform a clustering of the entities on the basis of all the information available, assuming that differentiation between the groups is to be with respect to the total information. Therefore in order to cluster the genotypes in the type of experiments described above, it would be desirable to consider a combination of both of these two-way arrays as a single three-way array. This produces a genotype by attribute by environment ($G \times A \times E$) matrix which is a three-mode three-way data set. A clustering technique is required to group one of these modes (the genotypes) on the basis of both of the other modes (attributes and environments). Such an approach is beneficial for two reasons. Firstly, significant genotype by environment interaction is almost always present, and it should be considered in the identification of groups of genotypes for which a general behavioral description is required. Secondly, it might be expected that the correlations between the attributes would differ across these groups of genotypes, and a single attribute analysis would not provide any information on this.

There has been some discussion in the literature on combining attributes to produce a biologically meaningful measure. For instance, seed yield and seed protein percentage could be combined to form a new attribute called seed protein yield. Selection indices (Smith, 1936 and Manning, 1956) are another way of combining attributes into a single measure. In many cases, however, no obvious or appropriate measure is available. Thus, if the clustering is to be based on the information inherent in all the attributes measured in each environment, then the proposed technique must be able to handle three-mode three-way data. Variable reduction techniques, such as principal component analysis or generalized canonical correlation, can be used to convert the data set to a two-way array, and thereby permit analysis with a conventional clustering technique. However, Chang (1983) showed that the practice of applying principal components to reduce the dimension of the data before clustering is not justified in general. By considering a mixture of two normal distributions with a common covariance matrix, he showed that the components with the larger eigenvalues do not necessarily provide more separation between the groups.

Carroll and Arabie (1983) devised a method for nonhierarchical overlapping clustering called INDCLUS for the case of three-way proximity data. Data in the form of a three-mode three-way array are first converted, using a similarity measure, to two-mode three-way data. This produces a matrix of entity by entity proximity measures for each of a number of individual subjects or data sources. Carroll, Clark and DeSarbo (1984) developed a new methodology called INDTREES for fitting a hierarchical tree structure to obtain a discrete network representation of such proximity data. The individual differences generalization is one in which individuals, for example, are assumed to base their judgements on the same family of trees, but are allowed to have different node heights and/or branch lengths. Their method minimizes the sum of the squared differences between fitted and observed dissimilarity between entities within a data source, while satisfying the ultrametric inequality. That is, given two disjoint clusters, all distances between entities in the same cluster are smaller than distances between entities in different clusters, and these between cluster distances are equal (Carroll, Clark and DeSarbo, 1984). It would be possible to apply such methods to the present problem by calculating a dissimilarity measure between genotypes within each environment and considering each environment as an individual data source. Squared Euclidean distance could be used as the proximity measure, although other choices are available.

Recently, DeSarbo, Carroll, Clark and Green (1984) proposed a new clustering method, called SYNCLUS, for clustering entities on which a battery of variables has been measured. It is an algorithm for $k$-means clustering using a weighted mean squared, stress-like measure, and can be generalized to handle three-way data. SYNCLUS can be applied in those situations where it is appropriate to put prior weighting on particular batteries of variables, and then allow the clustering procedure to weight the variables within these batteries according to their relative importance to the clustering. With respect to the present problem, the entities would be the genotypes and the attributes measured in each environment would be the variables in each battery. It is possible that the genotype by environment interaction might be expressed in the SYNCLUS model by different weightings on the attributes in each environment, but it might not be straightforward to interpret these weightings in terms of the interaction. A method of clustering which incorporates this genotype by environment interaction directly into the underlying model is desirable.

In analysing such three-way data, Basford (1982) considered a multidimensional scaling (MDS) approach to obtain a spatial representation in a low dimensional space. The relative proximity of the points (genotypes in this instance) in this space was then used as an indication of similarity of response pattern. Kruskal (1977), Whitmore and Harner (1980) and Morgan (1981) noted that, in general, a good overall picture is obtained using multidimensional scaling, but that it is not very sensitive to local features of the arrangement. Ramsay (1982) reviewed the statistical problems associated with multidimensional scaling, and in the subsequent discussion, the exploratory nature of this graphical technique was stressed. It should be clear, therefore, that multidimensional scaling is not a competing technique but rather a complementary one to clustering as previously stated by Kruskal (1977). An account of the technique of multidimensional scaling may be found in Shepard (1962a and b) and Kruskal (1964a and b). Kruskal and Wish (1978) gave a particularly clear exposition of the technique, illustrating its wide usage in the social and behavioral sciences as a descriptive model for elucidating data patterns.

One way of developing a clustering technique appropriate to three-way data is to fit a normal mixture model as proposed by Basford and McLachlan (1985c). Their proposal is to be described in the next section, followed by an example concerning some soybean data as analyzed by them. Note that although the problem has been cast in the framework of multiattribute genotype responses across environments, this approach is applicable to other three-way data sets. One example would be an investigation aimed at grouping individuals based on their responses to different tests, say manual, intellectual and memory tests, taken under several blood alcohol levels.

## 7.2 FITTING A NORMAL MIXTURE MODEL TO THREE-WAY DATA

The fitting of a normal mixture model to three-way data is presented here in the context of clustering genotypes on the basis of genotype by attribute by environment data. The observation vector $\mathbf{x}_j$ $(j = 1, \ldots, n)$ contains the multiattribute responses of the $j$th genotype in all $r$ environments, and is given by

$$\mathbf{x}_j = (\mathbf{x}'_{j1}, \ldots, \mathbf{x}'_{jr})',$$

where $\mathbf{x}_{jk}$ is a vector of length $p$ giving the response of genotype $j$ in environment $k$ for each of the same $p$ attributes measured in each environment $(k = 1, \ldots, r)$. The vectors $\mathbf{x}_{jk}$ $(j = 1, \ldots, n; k = 1, \ldots, r)$ are taken to be independently distributed. Under the mixture model proposed by Basford and McLachlan (1985c), it is assumed that each genotype belongs to one of $g$ possible groups $G_1, \ldots, G_g$ in proportions $\pi_1, \ldots, \pi_g$ respectively, so that in a given environment $k$,

$$\mathbf{x}_{jk} \sim N(\mu_{ik}, \Sigma_i) \quad \text{in } G_i \text{ with prob. } \pi_i \quad (i = 1, \ldots, g). \tag{7.2.1}$$

The within group covariance matrix $\boldsymbol{\Sigma}_i$ is taken not to depend on the environment. This model covers the general situation where there may be some interaction between genotypes and environments; indeed, in the example to be considered later there is a highly significant genotype by environment interaction for each attribute.

From (7.2.1), the density of the full observation vector $\mathbf{x}_j$ conditional on the $j$th genotype belonging to $G_i$ is equal to

$$f_i(\mathbf{x}_j; \theta) = (2\pi)^{-rp/2} |\Sigma_i|^{-r/2} \times \exp\left\{ -\tfrac{1}{2} \sum_{k=1}^{r} (\mathbf{x}_{jk} - \boldsymbol{\mu}_{ik})' \Sigma_i^{-1} (\mathbf{x}_{jk} - \boldsymbol{\mu}_{ik}) \right\} \tag{7.2.2}$$

for $i = 1, \ldots, g$, where the vector $\boldsymbol{\theta}$ of unknown group parameters contains the elements of $\boldsymbol{\mu}_{ik}$ $(i = 1, \ldots, g, ; k = 1, \ldots r)$ and the distinct elements of $\boldsymbol{\Sigma}_i$ $(i = 1, \ldots, g)$. The problem therefore is to fit a mixture of normal densities given by (7.2.2) to the independent data $\mathbf{x}_1, \ldots, \mathbf{x}_n$.

The likelihood estimate of $\phi = (\pi', \theta')'$ satisfies

$$\hat{\pi}_i = \sum_j \hat{\tau}_{ij}/n,$$

$$\hat{\boldsymbol{\mu}}_{ik} = \sum_j \hat{\tau}_{ij}\mathbf{x}_{jk}/n\hat{\pi}_i, \tag{7.2.3}$$

and

$$\hat{\boldsymbol{\Sigma}}_i = \sum_{jk} \hat{\tau}_{ij}(\mathbf{x}_{jk} - \hat{\boldsymbol{\mu}}_{ik})(\mathbf{x}_{jk} - \hat{\boldsymbol{\mu}}_{ik})'/nr\hat{\pi}_i$$

for $i = 1, \ldots, g$. The posterior probabilities $\tau_{ij}$ have the same form as before (see equation (1.4.3)), where now $f_i(\mathbf{x}_j; \theta)$ is given by (7.2.2).

The computation of the likelihood estimate of $\phi$ is facilitated by identifying these equations with the application of the EM algorithm of Dempster, Laird and Rubin (1977), as described in a general context in Section 1.6; a FORTRAN listing is given in the Appendix. Because of the amount of computation and possible difficulties involved in fitting a mixture model to three-way data, it is suggested that the number of attributes $p$ considered simultaneously be kept at a manageable level. Since the within group covariance matrix is allowed to be different for each group, each data point gives rise to a singularity in the likelihood on the edge of the parameter space, as noted in Section 2.1 for two-way data. Past experience in fitting this mixture model to genotype by attribute by environment data has shown that one genotype can be so distinct from the others that it is difficult to avoid the convergence of the EM algorithm to a singularity in the likelihood, corresponding to the clustering which has the distinct genotype on its own. However, with this genotype having been identified as distinct from the others, the mixture model can be fitted to the remaining genotypes.

Note from (7.2.3) that $\hat{\boldsymbol{\mu}}_{ik}$ can be written as

$$\hat{\boldsymbol{\mu}}_{ik} = \hat{\boldsymbol{\mu}}_i + \sum_{j=1}^{n} \hat{\tau}_{ij}(\mathbf{x}_{jk} - \bar{\mathbf{x}}_j)/n\hat{\pi}_i \qquad (i = 1, \ldots, g), \tag{7.2.4}$$

where

$$\bar{\mathbf{x}}_j = \sum_{k=1}^{r} \bar{\mathbf{x}}_{jk}/r$$

and where

$$\hat{\boldsymbol{\mu}}_i = \sum_{k=1}^{r} \hat{\boldsymbol{\mu}}_{ik}/r \tag{7.2.5}$$

is the estimate of the $i$th group mean over all environments. The second term on the right-hand side of (7.2.4) can be interpreted as an estimate of the sum of the effect of the $k$th environment and the interaction between the $i$th group and the $k$th environment.

## 7.3 CLUSTERING OF SOYBEAN DATA

To demonstrate the mixture likelihood approach to the clustering of three-way data, Basford and McLachlan (1985c) considered a soybean data set which had been discussed in the literature before and for which the adaptation of the genotypes is well known . This permitted some judgement on the usefulness of this method of clustering. Mungomery, Shorter and Byth (1974) first reported the experiment from which this set was collected, and it was subsequently analysed by Shorter, Byth and Mungomery (1977) and Basford (1982). Fifty-eight soybean lines, whose origin and maturity details are shown in Table 36, were evaluated at four locations in south-eastern Queensland in 1970 and 1971. The locations, Redland Bay, Lawes, Brookstead and Nambour, were within 150 kilometers of Brisbane, and covered a wide range of climatic, cultural and edaphic conditions. The experiment was a randomized complete block design with two replicates in each environment. Chemical and agronomic attributes were observed, including seed yield ($kg/ha$), plant height (cm), lodging (rating scale 1–5), seed size ($g/100$ seeds), seed protein percentage and seed oil percentage.

Because the pattern of group response is perhaps best interpreted by graphs of expected response for each attribute, it was decided to illustrate this method of clustering for $p = 2$ attributes. As each attribute was observed in four sites in two successive years, there were eight effective environments. There were two replications in each environment and, as with the analyses of Mungomery, Shorter and Byth (1974), the basic data set $\mathbf{x}_{jk}$ was taken to be the mean response over the replicates in each environment. The two attributes considered are seed yield and seed protein

**TABLE 36** Origin and maturity of soybean lines tested across four locations in each of two years

| Line No. | Name | Origin | Relative maturity |
|---|---|---|---|
| 1–40 | | LS* | Mid-very late (8–11)† |
| 41 | CPI 15939 Avoyelles | Tanzania | Late-mid (9) |
| 42 | CPI 15948 Hernon 49 | Tanzania | Late-mid (9) |
| 43 | CPI 17192 Mamloxi | Nigeria | Very late (11) |
| 44 | Dorman | U.S.A. | Early (5) |
| 45 | Hampton | U.S.A. | Mid (8) |
| 46 | Hill | U.S.A. | Early (5) |
| 47 | Jackson | U.S.A. | Early-mid (7) |
| 48 | Leslie | U.S.A. | Mid (8) |
| 49 | Semstar | Local cultivar | Mid-late (8) |
| 50 | Wills | U.S.A. | Mid (8) |
| 51 | CPI 26673 | Morocco | Very early (3) |
| 52 | CPI 26671 | Morocco | Very early (3) |
| 53 | Bragg | U.S.A. | Mid (7) |
| 54 | Delmar | U.S.A. | Early (4) |
| 55 | Lee | U.S.A | Early-mid (6) |
| 56 | Hood | U.S.A. | Early-mid (6) |
| 57 | Ogden | U.S.A. | Early-mid (6) |
| 58 | Wayne | U.S.A. | Very early (3) |

Adapted from Mungomery, Shorter and Byth (1974).

*LS, local selections from Mamloxi (CPI 17192) × Avoyelles (CPI 15939).

†Number in parenthesis is U.S. maturity group classification or estimated equivalent.

percentage. Mungomery, Shorter and Byth (1974) reported a clustering of the soybean lines using each of these separately.

The normal mixture model (7.2.1) was fitted to the soybean data set to obtain groups of genotypes within which there were similar patterns of performance. As recommended earlier, several starting values were used for each value of $g$, the specified number of underlying groups. An initial grouping of the genotypes can be obtained by focusing attention on a single attribute and using the corresponding analysis of variance table and subsequent multiple comparisons of genotype means. Alternatively, initial groupings can be obtained by using the results of other clustering techniques applied to the genotype by environment data for a single attribute. Both of these methods were tried here.

For a given $g$ ($g = 1$ to 9), the value of the log likelihood $L(\hat{\phi})$ is listed in Table 37, along with the $P$-value for the likelihood ratio test of $g$ versus

$g+1$ groups, computed according to the approximation (1.10.2), equivalent to $\chi^2_{38}$ here, for the null distribution of $-2\log\lambda$. It can be seen that the likelihood increases substantially as $g$ increases from one, but flattens out after $g = 7$. On the choice of $g$, the approximate $P$-values in Table 37 indicate $g = 7$, as the smallest value of $g$ compatible with the data. To provide further evidence in favor of this value of $g$, 19 bootstrap replications of $-2\log\lambda$ were generated for the test of $g = 7$ versus $g = 8$, in the same way as for the two-way data sets considered in previous chapters. As the value of 42.6 for $-2\log\lambda$ from the original data exceeds only 12 of the 19 bootstrap values generated for this statistic, the null hypothesis of $g = 7$ groups cannot be rejected at any reasonable significance level.

A clustering of the geneotypes is achieved by assigning each genotype to the group to which it has the highest estimated posterior probability of belonging. This clustering of the soybean lines into $g = 7$ groups is given in Table 38. As the smallest value of the maximum of the estimated posterior probabilities over the seven groups for a given genotype is 0.92, it would appear that the genotypes can be clustered into seven groups with a high degree of certainty under the normal mixture model adopted. The estimated group means, calculated from (7.2.5), for seed yield and protein percentage are reported in Table 39, along with the estimated correlation between these attributes in each group. The overall sample means and correlation are listed too.

**TABLE 37** Log likelihood and $P$-value for likelihood ratio test of $g$ versus $g+1$ groups under a normal mixture model

| $g$ | $L(\hat{\phi})$ | $P$ |
|---|---|---|
| 1 | −1468.4 | 0.000 |
| 2 | −1282.7 | 0.000 |
| 3 | −1207.5 | 0.000 |
| 4 | −1150.5 | 0.000 |
| 5 | −1099.0 | 0.002 |
| 6 | −1064.7 | 0.017 |
| 7 | −1035.3 | 0.280 |
| 8 | −1014.0 | 0.999 |

**TABLE 38** Clustering of soybean lines obtained by the mixture likelihood approach

| Group | Line numbers |
|---|---|
| $G_1$ | 51, 52, 58 |
| $G_2$ | 44, 46, 54 |
| $G_3$ | 45, 47, 48, 49, 50, 53, 55, 56, 57 |
| $G_4$ | 3, 4, 5, 6, 7, 8, 9, 10, 25 |
| $G_5$ | 1, 2, 14, 15, 16, 28, 31, 34, 35 |
| $G_6$ | 24, 26, 27, 32, 33, 38, 39, 41, 42 |
| $G_7$ | 11, 12, 13, 17, 18, 19, 20, 21, 22, 23, 29, 30, 36, 37, 40, 43 |

Source: Basford and McLachlan (1985c).

The normal mixture model fitted above has an arbitrary covariance matrix $\Sigma_i$ within each group $G_i$. However, in some instances it may be reasonable to take $\Sigma_i$ to be diagonal, which corresponds directly to the usual conditional independence factor model (Aitkin, Anderson and Hinde, 1981). That is, the observed correlations between the attributes result from the clustered nature of the sample, and that within the underlying groups, the attributes are independent. The present example appeared to show some support for this proposition, as the estimated correlation between yield and protein percentage within a group (see Table 39) was generally quite small.

An analysis of the data under the restricted model of a diagonal covari-

**TABLE 39** Estimated mean effect for each attribute and correlation within the groups

| Group | Mean yield (kg/ha) | Mean protein percentage | Correlation |
|---|---|---|---|
| $G_1$ | 1451.4 | 39.5 | −0.47 |
| $G_2$ | 2227.0 | 38.1 | 0.07 |
| $G_3$ | 2879.2 | 38.9 | −0.24 |
| $G_4$ | 2206.2 | 38.1 | 0.05 |
| $G_5$ | 1899.1 | 40.1 | −0.13 |
| $G_6$ | 2191.7 | 41.0 | −0.04 |
| $G_7$ | 1566.3 | 42.7 | −0.08 |
| Overall | 2047.4 | 40.3 | −0.34 |

Adapted from Basford and McLachlan (1985c).

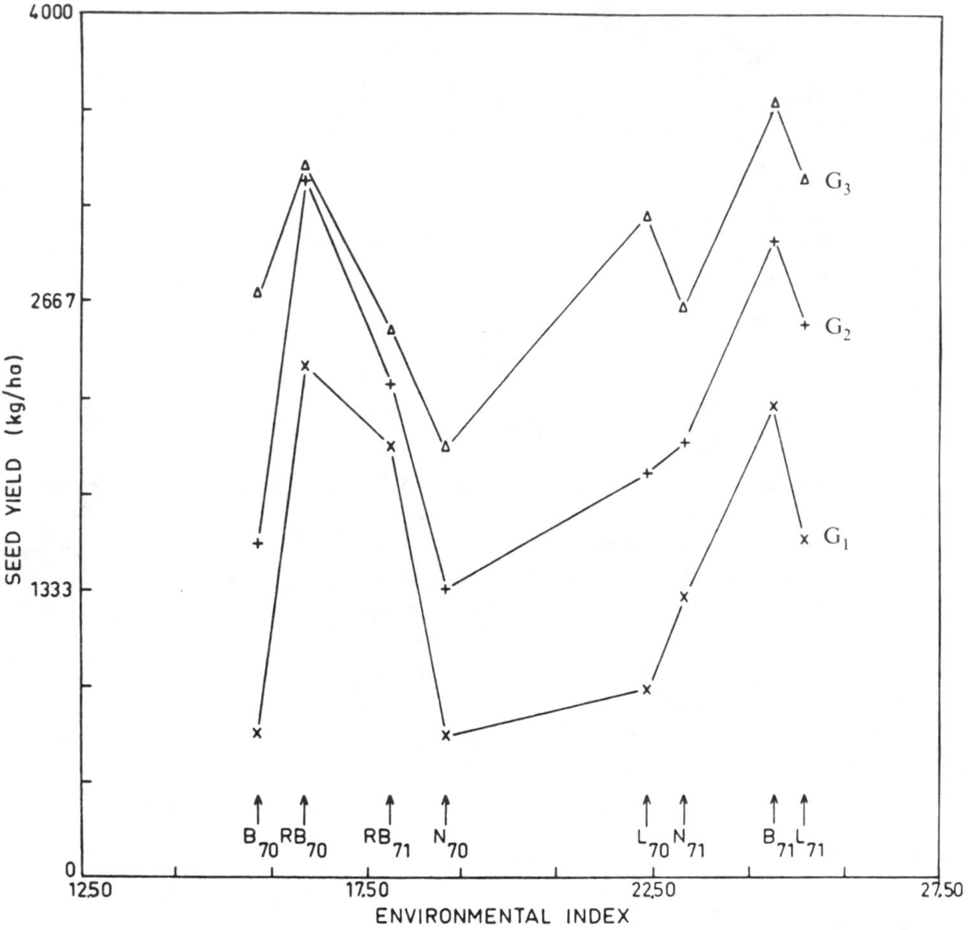


**FIGURE 7** Plot of estimated expected seed yield in each of groups $G_1$, $G_2$, and $G_3$ against each environment mean over all genotypes. (From Basford and McLachlan, 1985c.)

ance matrix within each group was undertaken, and resulted in a different clustering of the genotypes. A five group rather than a seven group description of the data set was concluded to be adequate, although group composition showed considerable resemblance to part of that displayed in Table 38.

It has been seen that by fitting the normal mixture model (7.2.1) to this three-way data set, a brief summary of the response patterns is obtained in

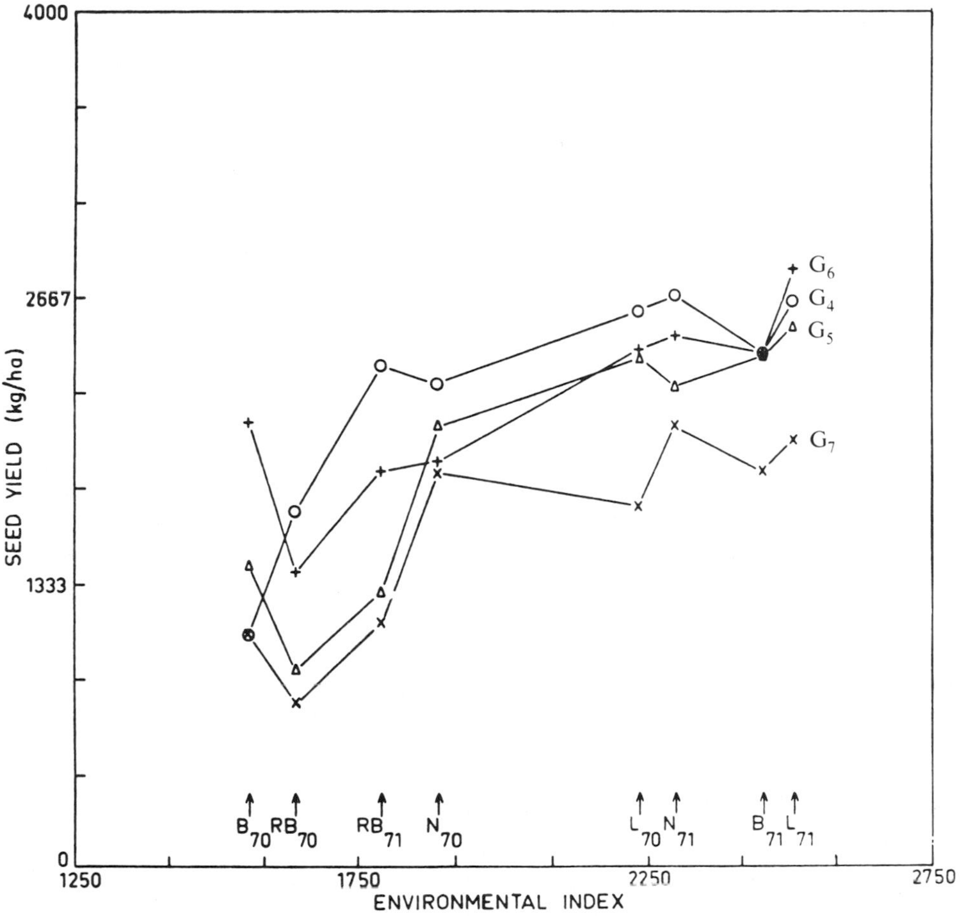


**FIGURE 8** Plot of estimated expected seed yield in each of groups $G_4$, $G_5$, $G_6$, and $G_7$ against each environment mean over all genotypes. (From Basford and McLachlan, 1985c.)

terms of the groups $G_1$ to $G_7$. To understand and explain the differences between the groups, it is beneficial here to consider the pattern of group responses across environments, as exhibited in Figures 7 to 10. In these figures, the estimated expected response $\hat{\mu}_{ik}$, for the $i$th genotype group in the $k$th environment is plotted against the $k$th environment mean over all genotypes for seed yield and protein percentage, respectively. In each case, the horizontal axis is an index of increasing environmental response

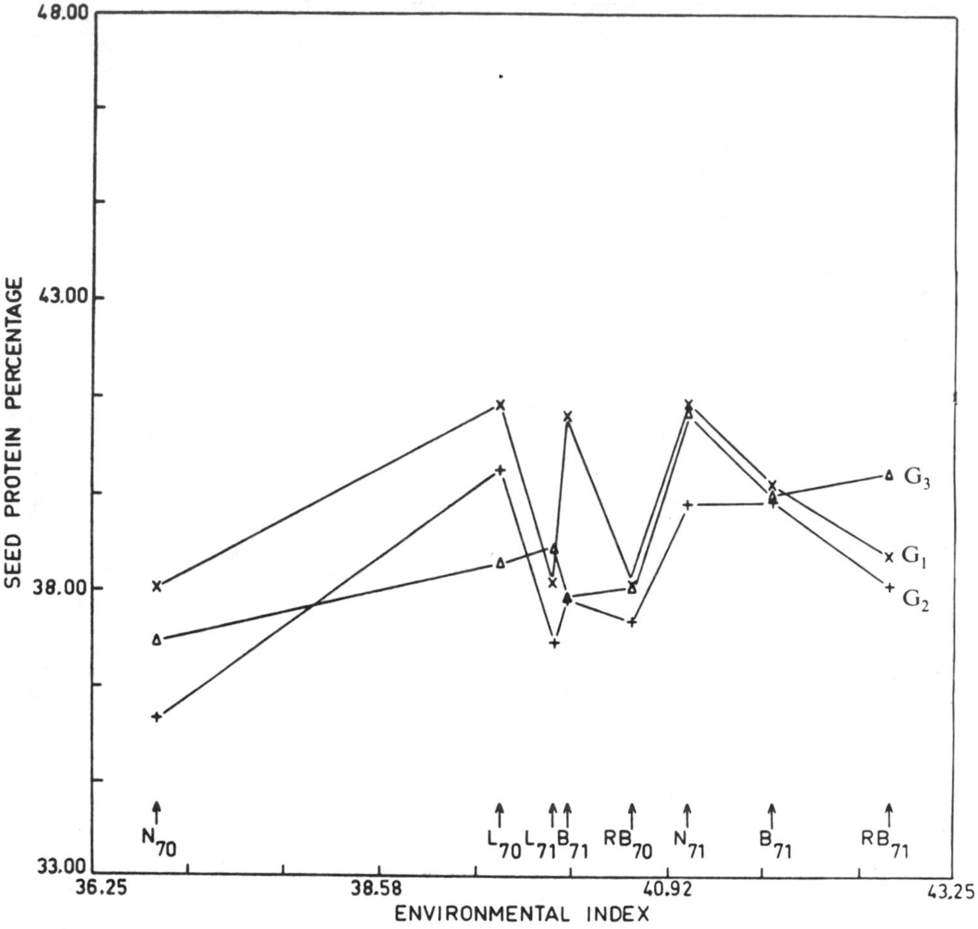


**FIGURE 9** Plot of estimated expected seed protein percentage in each of groups $G_1$, $G_2$ and $G_3$ against each environment mean over all genotypes. (From Basford and McLachlan, 1985c.)

for that attribute. The environments are denoted by the site initials and year subscript; for example, $RB_{70}$ refers to Redland Bay in 1970.

From an inspection of Figures 7 to 10, it is apparent that on the basis of response pattern, these groups can be viewed as two subsets: $G^{(1)}$ containing groups $G_1$ to $G_3$ and $G^{(2)}$ containing groups $G_4$ to $G_7$. The patterns are particularly alike within subset $G^{(1)}$ for yield and within sub-

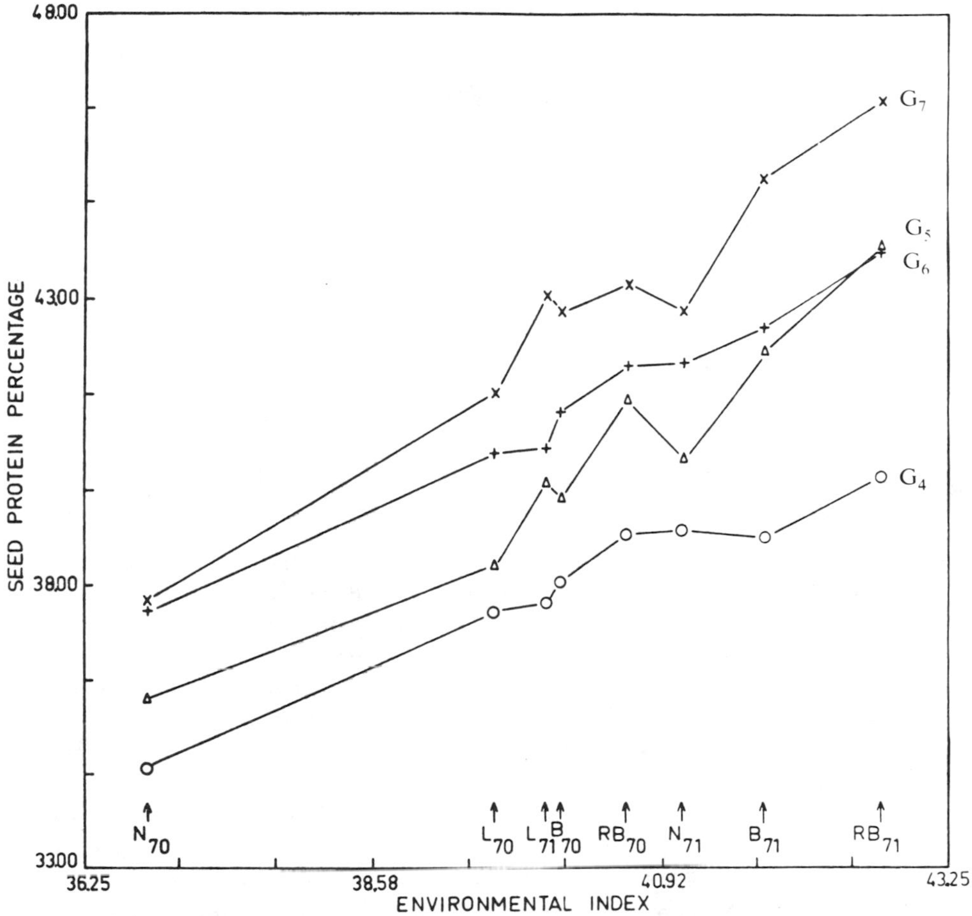


**FIGURE 10** Plot of estimated expected seed protein percentage in each of groups $G_4$, $G_5$, $G_6$ and $G_7$ against each environment mean over all genotypes. (From Basford and McLachlan, 1985c.)

set $G^{(2)}$ for protein percentage. Subset $G^{(1)}$ contains those lines with early to mid maturity group classification or estimated equivalent, while subset $G^{(2)}$ contains line 41, 43, all progeny from their cross and only one other line, 42, which is also late maturing. These patterns are discussed in more detail in Basford and McLachlan (1985c).

The mixture likelihood approach can be applied for $p > 2$ attributes, although the number of parameters in the model (7.2.1) increases sharply as $p$ increases, greatly compounding the problems with multiple maxima. A clustering of the genotypes using the responses of all six attributes at each location in each year was undertaken according to the mixture approach. The same number of underlying groups was obtained with only slightly different group composition to that obtained using just seed yield and seed protein percentage. It resulted in a grouping of line 42 with lines 44 and 54 in $G_2$ of subset $G^{(1)}$, whereas it was put in $G_6$ in subset $G^{(2)}$ in the analysis based on yield and protein percentage only. All other differences between the groups obtained from using all six attributes rather than just these two were rearrangements within each of the subsets $G^{(1)}$ and $G^{(2)}$; that is, line 46 went to $G_3$ from $G_2$, line 38 went to $G_5$ from $G_6$ and line 40 went to $G_6$ from $G_7$.

## 7.4 MULTIDIMENSIONAL SCALING APPROACH TO THE ANALYSIS OF SOYBEAN DATA

As remarked in the introduction to this chapter, multidimensional scaling is not a competing but rather a complementary technique to clustering. Proximities among the entities under consideration are used to determine a spatial representation consisting of a geometric configuration of points, as on a map, where each point corresponds to one of the entities. Three-way multidimensional scaling uses several matrices of proximities, constituting a three-way array, and allows for large systematic differences among the matrices (Tucker and Messick, 1963; Carroll and Chang, 1970). In applying multidimensional scaling to the soybean data considered in the previous section, it is postulated that an underlying pattern of genotype performance, as measured by an array of attributes across environments, exists, and that this pattern reflects both main and interaction effects. Under this three-way model, the positions of the genotypes, as determined by an environment, may vary by a change in the relative importance of the conceptual axes of the underlying space. In this way, each environment may elicit different responses from the genotypes. The object of the analysis is to identify the underlying dimensions and to investigate the placing and relationships of the genotypes in that space.

In order to formally define the individual differences model (Carroll and Chang, 1970), a set of $q$ dimensions is assumed to underlie the genotype space common to all environments. The dissimilarity between the genotypes for each environment is assumed to relate in a simple way to a modified Euclidean distance model; that is, the distance between genotypes $i$ and $j$ for environment $k$ is given by

$$d_{ijk} = \sqrt{\sum_{t=1}^{q} u_{kt}(x_{it} - x_{jt})^2}, \qquad (7.4.1)$$

where $x_{it}$ is the projection of the $i$th genotype on the $t$th dimension and $u_{kt}$ is the weight placed by the $k$th environment on the $t$th dimension. The weights represent the relative importance of the underlying dimensions for the individual environments. This method exploits the differences among the environments to give unique axes which, provided the model is true, correspond to meaningful dimensions. In fitting the model, the aim is to match the distance given in equation (7.4.1) to the observed distance between the genotypes within an environment.

The program ALSCAL (Takane, Young and de Leeuw, 1977) was used, as it fits the Euclidean model in a metric or nonmetric procedure and contains some refinements to the implementation of Carroll and Chang (1970). The nonmetric option is less restrictive than the metric one, which may use an inappropriate assumption about the relationship between observed distances between genotypes within an environment and the distances given in equation (7.4.1). However, the nonmetric option offers less resistance to degenerate solutions and locally optimal solutions than some forms of metric scaling. Kruskal and Wish (1978) stated that "The use of metric scaling instead of nonmetric scaling has very little effect on the configuration in most cases," and this was verified to be true with this data set. As no problems with degeneracy were encountered, the nonmetric option was used. Takane, Young and de Leeuw (1977) reported that (i) Monte Carlo studies showed ALSCAL to be very robust and most resistant to local minima, and (ii) the procedure for fitting the model belongs to a class of numerical procedures termed alternating least squares which have the desirable property of being necessarily convergent.

For ease of interpretation, it is desirable to express the configuration in a low, two or three, dimensional space. Representation of data in a reduced space, however, inevitably results in some loss of information if the underlying space is of a higher dimension. To assess the adequacy of the model, Young and Lewyckyj (1980) suggested the average squared

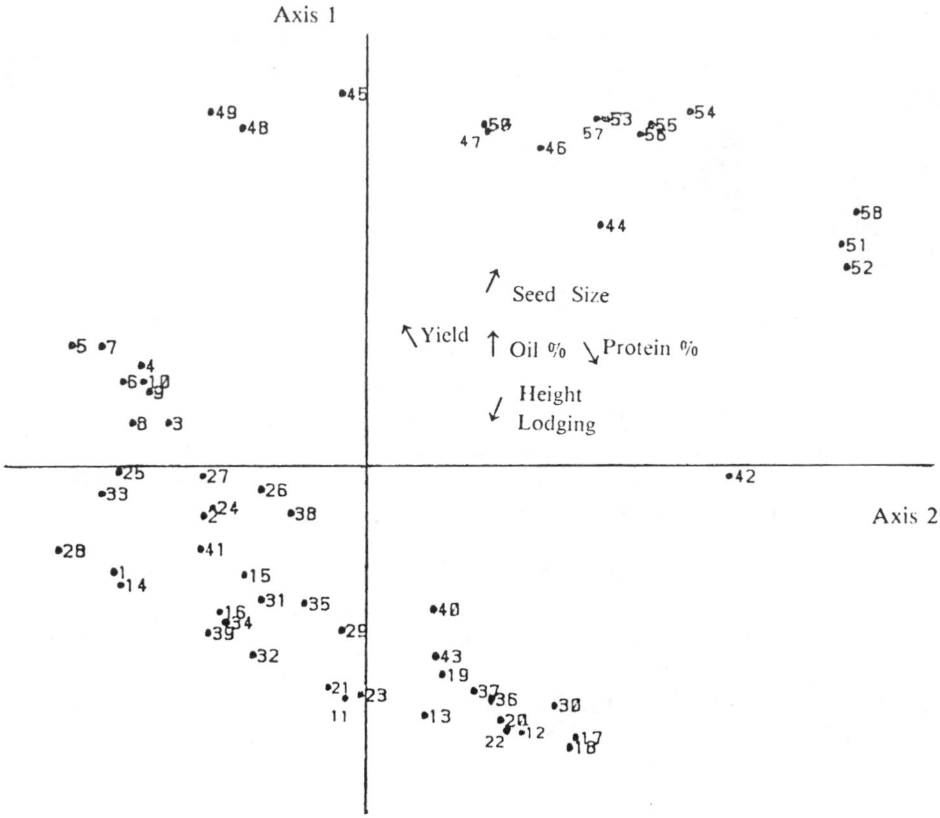


**FIGURE 11** Two dimensional spatial representation from MDS analysis of 58 soybean lines; arrows indicate the direction of increasing value of the six measured attributes used in the analysis. (From Basford, 1982.)

correlation between distances, that is,

$$\sum_{k=1}^{r} C_k^2/r, \tag{7.4.2}$$

where $r$ is the number of environments and $C_k$ is the correlation between observed and fitted distances between the genotypes for environment $k$. As $C_k^2$ reflects the variance accounted for in the $k$th environment, (7.4.2) indicates the average proportion of variation accounted for by the multidimensional scaling model. When different measurement types are involved the data need to be scaled. Here the usual standardization was chosen using attribute means and standard deviations over all environments, rather than within each one, so as to retain the differences due to the within environment variation.

The two-dimensional spatial representation of the soybean lines accounted for an average of 79.4% of the variation in distances, using all six attributes to characterize the genotypes in the same eight environments. This spatial representation is displayed in Figure 11. As discussed in more detail in Basford (1982), each environment placed greater weight on the first axis, reflecting its overriding importance in determining the spatial position of the genotypes in the reduced space.

As remarked earlier, multidimensional scaling and clustering have a strong complementary relationship. For example, the two-dimensional scaling configuration of the genotypes in Figure 11 can be used to portray the clustering produced by the mixture likelihood approach by drawing, say, loops to represent the clusters. For the clustering of the soybean lines into $g = 7$ groups as given in Table 38, the separation between the two subsets $G^{(1)}$ and $G^{(2)}$ of these seven groups is clearly evident from Figure 11, apart for line 42 which appears quite distinct from the other lines. Several observations of Basford (1982) arising from the multidimensional scaling analysis of the soybean data are in agreement with the results of the mixture analysis. For example, Nambour had similar weight ratios for the two axes of the spatial representation for each of 1970 and 1971, and this location showed consistency with respect to the spread of the groups (Figures 7 to 10).

# References

Agha, M. and Ibrahim, M.T. (1984). Algorithm AS 203. Maximum likelihood estimation of mixtures of distributions. *Appl. Statist.* **33**, 327–332.

Aitchison, J. and Aitken, C.G.G. (1976). Multivariate binary discrimination by the kernel method. *Biometrika* **63**, 413–420.

Aitchison, J. and Dunsmore, I.R. (1975). *Statistical Prediction Analysis.* Cambridge: Cambridge University Press.

Aitchison, J., Habbema, J.D.F. and Kay, J.W. (1977). A critical comparison of two methods of statistical discrimination. *Appl. Statist.* **26**, 15–25.

Aitkin, M. (1980a). Mixture applications of the EM algorithm in GLIM. *Compstat 1980, Proc. Computational Statistics.* Vienna: Physica-Verlag, pp. 537–541.

Aitkin, M. (1980b). Contribution to the discussion of paper by D.J. Bartholomew. *J. R. Statist. Soc. B* **42**, 312–314.

Aitkin, M., Anderson, D. and Hinde, J. (1981). Statistical modelling of data on teaching styles (with discussion). *J. R. Statist. Soc. A* **144**, 419–461.

Aitkin, M. and Rubin, D.B. (1985). Estimation and hypothesis testing in finite mixture models. *J. R. Statist. Soc. B* **47**, 67–75.

Aitkin, M. and Tunnicliffe Wilson, G. (1980). Mixture models, outliers, and the EM algorithm. *Technometrics* **22**, 325–331.

Amoh, R.K. (1985). Estimation of a discriminant function from a mixture of two inverse Gaussian distributions when sample size is small. *J. Statist. Comput. Simul.* **20**, 275–286.

Anderberg, M.R.C. (1973). *Cluster Analysis for Applications.* New York: Academic Press.

Anderson, J.A. (1972). Separate sample logistic discrimination. *Biometrika* **59**, 19–35.

Anderson, J.A. (1979). Multivariate logistic compounds. *Biometrika* **66**, 7–16.

Anderson, J.J. (1985). Normal mixtures and the number of clusters problem. *Comput. Statist. Quarterly* **2**, 3–14.

Anderson, R.L. and Bancroft, T.A. (1952). *Statistical Theory in Research.* New York: McGraw-Hill.

Anderson, T.W. (1951). Classification by multivariate analysis. *Psychometrika* **16**, 31–50.

Anderson, T.W. (1958). *An Introduction to Multivariate Statistical Analysis.* First Edition. New York: Wiley.

Anderson, T.W. (1984). *An Introduction to Multivariate Statistical Analysis.* Second Edition. New York: Wiley.

Andrews, D.F. (1972). Plots of high dimensional data. *Biometrics* **28**, 125–136.

Barnett, V. (1983). Contribution to the discussion of paper by R.J. Beckman and R.D. Cook. *Technometrics* **25**, 150–152.

Barnett, V. and Lewis, T. (1978). *Outliers in Statistical Data.* New York: Wiley.

Bartlett, M.S. (1938). Further aspects of multiple regression. *Proc. Camb. Phil. Soc.* **34**, 33–40.

Basford, K.E. (1982). The use of multidimensional scaling in analysing multiattribute genotype response across environments. *Aust. J. Agric. Res.* **33**, 473–480.

Basford, K.E. (1985). *Cluster Analysis Via Normal Mixture Models.* Unpublished Ph.D. thesis, University of Queensland.

Basford, K.E. and Horton, I.F. (1984). Approximating the likelihood for the clustering of treatment means with random block effects. *Comput. Statist. Quarterly* **1**, 115–121.

Basford, K.E. and McLachlan, G.J. (1983). On computational aspects associated with bias correction techniques in a cluster analysis context. *Proc. STATCOMP 83. Sydney: Statist. Soc. Austral. (Statist. Comput. Section)*, pp. 31–39.

Basford, K.E. and McLachlan, G.J. (1985a). Estimation of allocation rates in a cluster analysis context. *J. Amer. Statist. Assoc.* **80**, 286–293.

Basford, K.E. and McLachlan, G.J. (1985b). Cluster analysis in a randomized complete block design. *Commun. Statist.-Theor. Meth.* **14**, 451–463.

Basford, K.E. and McLachlan, G.J. (1985c). The mixture method of clustering applied to three-way data. *J. Classification* **2**, 109–125.

Basford, K.E. and McLachlan, G.J. (1985d). Likelihood estimation with normal mixture models. *Appl. Statist.* **34**, 282–289.

Bayne, C.K., Beauchamp, J.J., Begovich, C.L. and Kane, V.E. (1980). Monte Carlo comparisons of selected clustering procedures. *Pattern Recognition* **12**, 51–62.

Beckman, R.J. and Cook, R.D. (1983). Outlier...s(with discussion). *Technometrics* **25**, 119–163.

Begovich, C.L. and Kane, V.E. (1982). Estimating the number of groups and group membership using simulation cluster analysis. *Pattern Recognition* **15**, 335–342.

Behboodian, J. (1970). On a mixture of normal distributions. *Biometrika* **57**, 215–217.

Behboodian, J. (1972). Information matrix for a mixture of two normal distributions. *J. Statist. Comput. Simul.* **1**, 295–314.

Bera, A. and John, S. (1983). Tests for multivariate normality with Pearson alternatives. *Commun. Statist.-Theor. Meth.* **12**, 103–117.

Beran, R. (1984). Minimum distance procedures. In *Handbook of Statistics* (Vol. 4), P.R. Krishnaiah and P.K. Sen (Eds.). Amsterdam: North Holland, pp. 741–754.

Bezdek, J.C., Hathaway, R.J. and Huggins, V.J. (1985). Parametric estimation for normal mixtures. *Pattern Recognition Letters* **3**, 79–84.

Bhattacharya, C.G. (1967). A simple method for resolution of a distribution into its Gaussian components. *Biometrics* **23**, 115–135.

Binder, D.A. (1978). Bayesian cluster analysis. *Biometrika* **65**, 31–38.

Binder, D.A. (1981). Approximations to Bayesian clustering rules. *Biometrika* **68**, 275–285.

Blischke, W.R. (1978). Mixtures of distributions. In *International Encyclopedia of Statistics* (Vol. 1), W.H. Kruskal and J.M. Tanur (Eds.). New York: The Free Press, pp. 174–180.

Bock, H.H. (1985). On some significance tests in cluster analysis. *J. Classification* **2**, 77–108.

Boes, D.C. (1966). On the estimation of mixing distributions. *Ann. Math. Statist.* **37**, 177-188.

Box, G.E.P. and Cox, D.R. (1964). An analysis of transformations. *J. R. Statist. Soc. B* **26**, 211-252.

Box, G.E.P. and Tiao, G.C. (1968). A Bayesian approach to some outlier problems. *Biometrika* **55**, 119–129.

Boyles, R.A. (1983). On the convergence of the EM algorithm. *J. R. Statist. Soc. B* **45**, 47–50.

Bozdogan, H. and Sclove, S.L. (1984). Multi-sample cluster analysis using Akaike's information criterion. *Ann. Inst. Statist. Math.* **36**, 163–180.

Brownie, C., Habicht, J.–P. and Robson, D.S. (1983). An estimation procedure for the contaminated normal distributions arising in clinical chemistry. *J. Amer. Statist. Assoc.* **78**, 228–237.

Bryant, J.L. and Paulson, A.S. (1983). Estimation of mixing proportions via distance between characteristic functions. *Commun. Statist.-Theor. Meth.* **12**, 1009–1029.

Bryant, P. and Williamson, J.A. (1978). Asymptotic behaviour of classification maximum likelihood estimates. *Biometrika* **65**, 273–281.

Bryant, P.G. and Williamson, J.A. (1985). Maximum likelihood and classification: a comparison of three approaches. Faculty Working Paper No. 5, College of Business and Administration, University of Colorado, Denver.

Burt, R.L., Edye, L.A., Williams, W.T., Grof, B. and Nicholson, C.H.L. (1971). Numerical analysis of variation patterns in the genus *Stylosanthes* as an aid to plant introduction and assessment. *Aust. J. Agric. Res.* **22**, 737–757.

Butler, R.W. (1986). Predictive likelihood inference with applications (with discussion). *J. R. Statist. Soc. B* **48**, 1–38.

Byth, D.E., Eisemann, R.L. and DeLacy, I.H. (1976). Two-way pattern analysis of a large data set to evaluate genotypic adaptation. *Heredity* **37**, 215–230.

Cacoullos, T. (Ed.). (1973). *Discriminant Analysis and Applications.* New York: Academic Press.

Caliński, T. and Corsten, L.C.A. (1985). Clustering means in ANOVA by simultaneous testing. *Biometrics* **41**, 39–48.

Caliński, T. and Harabasz, J. (1974). A dendrite method for cluster analysis. *Commun. Statist.-Theor. Meth.* **3**, 1–27.

Campbell, N.A. (1980). Robust procedures in multivariate analysis. I: Robust covariance estimation. *Appl. Statist.* **29**, 231–237.

Campbell, N.A. (1984). Mixture models and atypical values. *Math. Geol.* **16**, 465–477.

Campbell, N.A. (1985). Mixture models—some extensions. Unpublished manuscript.

Carmer, S.G. and Lin, W.T. (1983). Type I error rates for divisive clustering methods for grouping means in analysis of variance. *Commun. Statist.-Simula. Computa.* **12**, 451–466.

Carroll, J.D. and Arabie, P. (1980). Multidimensional Scaling. *Ann. Rev. Psych.* **31**, 607–649.

Carroll, J.D. and Arabie, P. (1983). INDCLUS: An individual differences generalization of the ADCLUS model and the MAPCLUS algorithm. *Psychometrika* **48**, 157–169.

Carroll, J.D. and Chang, J.J. (1970). Analysis of individual differences in multidimensional scaling via an $N$-way generalization of Eckart-Young decomposition. *Psychometrika* **35**, 283–319.

Carroll, J.D., Clark, L.A. and DeSarbo, W.S. (1984). The representation of three-way proximity data by single and multiple tree structure models. *J. Classification* **1**, 25–74.

Cassie, R.M. (1954). Some uses of probability paper for the graphical analysis of polymodal frequency distributions. *Austral. J. Marine and Freshwater Res.* **5**, 513–522.

Chang, W.C. (1976). The effects of adding a variable in dissecting a mixture of two normal populations with a common covariance matrix. *Biometrika* **63**, 676–678.

Chang, W.C. (1979). Confidence interval estimation and transformation of data in a mixture of two multivariate normal distributions with any given large dimension. *Technometrics* **21**, 351–355.

Chang, W.C. (1983). On using principal components before separating a mixture of two multivariate normal distributions. *Appl. Statist.* **32**, 267–275.

Charlier, C.V.L. and Wicksell, S.D. (1924). On the dissection of frequency functions. *Arkiv för Mathematik, Astronomi och Fysik.* **Bd. 18**, No. 6.

Chhikara, R.S. and Register, D.T. (1979). A numerical classification method for partitioning of a large multidimensional mixed data set. *Technometrics* **21**, 531–537.

Choi, K. (1969a). Estimators for the parameters of a finite mixture of distributions. *Ann. Inst. Statist. Math.* **21**, 107–116.

Choi, K. (1969b). Empirical Bayes procedure for (pattern) classification with stochastic learning. *Ann. Inst. Statist. Math.* **21**, 117–125.

Choi, K. and Bulgren, W.G. (1968). An estimation procedure for mixtures of distributions. *J. R. Statist. Soc. B* **30**, 444–460.

Choi, S.C. (1979). Two-sample tests for compound distributions for homogeneity of mixing proportions. *Technometrics* **21**, 361–365.

Choi, S.C. (Ed.). (1986). *Statistical Methods of Discrimination and Classification—Advances in Theory and Applications.* New York: Pergamon Press.

Clifford, H.T. and Stephenson, W. (1975). *An Introduction to Numerical Classification.* New York: Academic Press.

Cohen, A.C. (1967). Estimation in mixtures of two normal distributions. *Technometrics* **9**, 15–28.

Collins, J.R. and Wiens, D.P. (1985). Minimax variance $M$-estimators in $\varepsilon$-contamination models. *Ann. Statist.* **13**, 1078–1096.

Cormack, R.M. (1971). A review of classification (with discussion). *J. R. Statist. Soc.* A **134**, 321–367.

Cox, D.R. (1966). Some procedures associated with the logistic qualitative response curve. In *Research Papers on Statistics: Festschrift for J. Neyman*, F.N. David (Ed.). New York: Wiley, pp. 55–71.

Cox, D.R. and Hinkley, D.V. (1974). *Theoretical Statistics.* London: Chapman and Hall.

Cox, D.R. and Small, N.J.H. (1978). Testing multivariate normality. *Biometrika* **65**, 263–272.

Cox, D.R. and Spjøtvoll, E. (1982). On partitioning means into groups. *Scand. J. Statist.* **9**, 147–152.

Cramér, H. (1946). *Mathematical Methods of Statistics.* Princeton: Princeton University Press.

Das Gupta, S. (1965). Optimum classification rules for classification into two multivariate normal populations. *Ann. Math. Statist.* **36**, 1174–1184.

Day, N.E. (1969). Estimating the components of a mixture of two normal distributions. *Biometrika* **56**, 463–474.

Day, N.E. and Kerridge, D.F. (1967). A general maximum likelihood discriminant. *Biometrics* **23**, 313–328.

Dempster, A.P., Laird, N.M. and Rubin, D.B. (1977). Maximum likelihood from incomplete data via the EM algorithm (with discussion). *J. R. Statist. Soc. B* **39**, 1–38.

DeSarbo, W.S., Carroll, J.D., Clark, L.A. and Green, P.E. (1984). Synthesized clustering: a method for amalgamating alternative clustering bases with differential weighting of variables. *Psychometrika* **49**, 57–78.

Desu, M.M. and Geisser, S. (1973). Methods and applications of equal-mean discrimination. In *Discriminant Analysis and Applications*, T. Cacoullos (Ed.). New York: Academic Press, pp. 139–159.

Devijver, P.A. and Kittler, J. (1982). *Pattern Recognition: A Statistical Approach.* London: Prentice Hall.

Devlin, S.J., Gnanadesikan, R. and Kettenring, J.R. (1981). Robust estimation of dispersion matrices and principal components. *J. Amer. Statist. Assoc.* **76**, 354–362.

Dick, N.P. and Bowden, D.C. (1973). Maximum likelihood estimation for mixtures of two normal distributions. *Biometrics* **29**, 781–790.

Do, K. and McLachlan, G.J. (1984). Estimation of mixing proportions: a case study. *Appl. Statist.* **33**, 134–140.

Duda, R.O. and Hart, P.E. (1973). *Pattern Classification and Scene Analysis.* New York: Wiley.

Duncan, D.B. (1955). Multiple range and multiple $F$ tests. *Biometrics* **11**, 1–42.

Dunn, C.L. (1982). Comparison of combinatoric and likelihood ratio procedures for classifying samples. *Commun. Statist.-Theor. Meth.* **11**, 2361–2377.

Dunn, C.L. and Smith, W.B. (1980). Combinatoric classification of multivariate normal variates. *Commun. Statist.-Theor. Meth.* **A9**, 1317–1340.

Dunn, C.L. and Smith, W.B. (1982). Normal combinatoric classification: the sample case. *Commun. Statist.-Theor. Meth.* **11**, 271–289.

Durairajan, T.M. and Kale, B.K. (1979). Locally most powerful test for the mixing proportion. *Sankhyā B* **41**, 91–100.

Durairajan, T.M. and Kale, B.K. (1982). Locally most powerful similar test for mixing proportion. *Sankhyā A* **44**, 153–161.

Duran, B.S. and Odell, P.L. (1974). *Cluster Analysis: A Survey.* Berlin: Springer-Verlag.

Edwards, A.W.F. and Cavalli-Sforza, L.L. (1965). A method for cluster analysis. *Biometrics* **21**, 362-375.

Efron, B. (1979). Bootstrap methods: another look at the jackknife. *Ann. Statist.* **7**, 1–26.

Efron, B. (1981a). Nonparametric estimates of standard error: the jackknife, the bootstrap and other methods. *Biometrika* **68**, 589–599.

Efron, B. (1981b). Nonparametric standard errors and confidence intervals (with discussion). *Canad. J. Statist.* **9**, 139–172.

Efron, B. (1982). *The Jackknife, the Bootstrap and Other Resampling Plans.* Philadelphia: SIAM.

Efron, B. (1983). Estimating the error rate of a prediction rule: improvement on cross-validation. *J. Amer. Statist. Assoc.* **78**, 316–331.

Efron, B. (1984). Better bootstrap confidence intervals. Technical Report No. 226. Stanford: Dept. of Statistics, Stanford University.

Efron, B. and Gong, G. (1983). A leisurely look at the bootstrap, the jackknife, and cross-validation. *Amer. Statistician* **37**, 36–48.

Efron, B. and Hinkley, D.V. (1978). Assessing the accuracy of the maximum likelihood estimator: observed versus expected Fisher information. *Biometrika* **65**, 457–487.

Efron, B. and Tibshirani, R. (1986). Bootstrap methods for standard errors, confidence intervals, and other measures of statistical accuracy (with discussion). *Statist. Science* **1**, 54–77.

Engelman, L. and Hartigan, J.A. (1969). Percentage points of a test for clusters. *J. Amer. Statist. Assoc.* **64**, 1647–1648.

Everitt, B.S. (1980). *Cluster Analysis*. Second Edition. London: Wiley-Halsted.

Everitt, B.S. (1981a). A Monte Carlo investigation of the likelihood ratio test for the number of components in a mixture of normal distributions. *Multivar. Behav. Res.* **16**, 171–180.

Everitt, B.S. (1981b). Contribution to the discussion of paper by M. Aitkin, D. Anderson and J. Hinde. *J. R. Statist. Soc.* A **144**, 457–458.

Everitt, B.S. (1984a). Maximum likelihood estimation of the parameters in a mixture of two univariate normal distributions; a comparison of different algorithms. *The Statistician* **33**, 205–215.

Everitt, B.S. (1984b). A note on parameter estimation for Lazarsfeld's latent class model using the EM algorithm. *Multivar. Behav. Res.* **19**, 79–89.

Everitt, B.S. (1985). Mixture distributions. In *Encyclopedia of Statistical Sciences* (Vol. 5), S. Kotz and N.L. Johnson (Eds.). New York: Wiley, pp. 559–569.

Everitt, B.S. and Hand, D.J. (1981). *Finite Mixture Distributions*. London: Chapman and Hall.

Fatti, L.P., Hawkins, D.M. and Raath, E.L. (1982). Discriminant analysis. In *Topics in Applied Multivariate Analysis*, D.M. Hawkins (Ed.). Cambridge: Cambridge University Press, pp. 1–71.

Fowlkes, E.B. (1979). Some methods for studying the mixture of two normal (lognormal) distributions. *J. Amer. Statist. Assoc.* **74**, 561–575.

Friedman, H.P. and Rubin, J. (1967). On some invariant criteria for grouping data. *J. Amer. Statist. Assoc.* **62**, 1159–1178.

Fryer, J.G. and Robertson, C.A. (1972). A comparison of some methods for estimating mixed normal distributions. *Biometrika* **59**, 639–648.

Fukunaga, K. (1972). *Introduction to Statistical Pattern Recognition*. New York: Academic Press.

Fukunaga, K. and Kessel, D.L. (1972). Application of optimum error-reject functions. *IEEE Trans. Inform. Theory* **IT–18**, 814–817.

Fukunaga, K. and Kessel, D.L. (1973). Nonparametric Bayes error estimation using unclassified samples. *IEEE Trans. Inform. Theory* **IT–19**, 434–440.

Gabriel, K.R. (1964). A procedure for testing the homogeneity of all sets of means in analysis of variance. *Biometrics* **20**, 459–477.

Ganesalingam, S. (1980). *On the Mixture Maximum Likelihood Approach to Estimation and Clustering*. Unpublished Ph.D. thesis, University of Queensland.

Ganesalingam, S. and McLachlan, G.J. (1978). The efficiency of a linear discriminant function based on unclassified initial samples. *Biometrika* **65**, 658–662.

Ganesalingam, S. and McLachlan, G.J. (1979a). Small sample results for a linear discriminant function estimated from a mixture of normal populations. *J. Statist. Comput. Simul.* **9**, 151–158.

Ganesalingam, S. and McLachlan, G.J. (1979b). A case study of two clustering methods based on maximum likelihood. *Statist. Neerlandica* **33**, 81–90.

Ganesalingam, S. and McLachlan, G.J. (1980a). A comparison of the mixture and classification approaches to cluster analysis. *Commun. Statist.-Theor. Meth.* **A9**, 923–933.

Ganesalingam, S. and McLachlan, G.J. (1980b). Error rate estimation on the basis of posterior probabilities. *Pattern Recognition* **12**, 405–413.

Ganesalingam, S. and McLachlan, G.J. (1981). Some efficiency results for the estimation of the mixing proportion in a mixture of two normal distributions. *Biometrics* **37**, 23–33.

Ghosh, J.K. and Sen, P.K. (1985). On the asymptotic performance of the log likelihood ratio statistic for the mixture model and related results. Proc. Berkeley Conference in Honor of Jerzy Neyman and Jack Kiefer (Vol. II), L.M. Le Cam and R.A. Olshen (Eds.). Monterey: Wadsworth, pp. 789–806.

Glick, N. (1978). Additive estimators for probabilities of correct classification. *Pattern Recognition* **10**, 211–222.

Gnanadesikan, R. (1977). *Methods for Statistical Data Analysis of Multivariate Observations.* New York: Wiley.

Goldstein, M. and Dillon, W.R. (1978). *Discrete Discriminant Analysis.* New York: Wiley.

Gong, G. (1986). Cross-validation, the jackknife, and the bootstrap: excess error estimation in forward logistic regression. *J. Amer. Statist. Assoc.* **81**, 108–113.

Goodall, C. (1983). *M*-estimators of location: an outline of the theory. In *Understanding Robust and Exploratory Data Analysis*, D.C. Hoaglin, F. Mosteller and J.W. Tukey (Eds.). New York: Wiley, pp. 339–403.

Gordon, A.D. (1981). *Classification.* London: Chapman and Hall.

Gordon, A.D. (1982). Some observations on methods of estimating the proportions of morphologically similar pollen types in fossil samples. *Can. J. Bot.* **60**, 1888–1894.

Gupta, S.S. and Huang, W.T. (1981). On mixtures of distributions: a survey and some new results on ranking and selection. *Sankhyā B* **43**, 245–290.

Guseman, L.F. and Walton, J.R. (1977). An application of linear feature selection to estimation of proportions. *Commun. Statist.-Theor. Meth.* **A6**, 611–617.

Guseman, L.F. and Walton, J.R. (1978). Methods for estimating proportions of convex combinations of normals using linear feature selections. *Commun. Statist.-Theor. Meth.* **A7**, 1439–1450.

Habbema, J.D.F., Hermans, J. and van den Broek, K. (1974). A stepwise discriminant analysis program using density estimation. *Compstat 1974, Proc. Computational Statistics.* Vienna: Physica-Verlag, pp. 101–110.

Hall, P. (1981). On the non-parametric estimation of mixture proportions. *J. R. Statist. Soc. B* **43**, 147–156.

Hall, P. (1986). Cross-validation in nonparametric density estimation. *Proc. XIII$^{th}$ Int. Biometric Conference*. Seattle: Biometric Society, 15 pp.

Hall, P. and Titterington, D.M. (1984). Efficient nonparametric estimation of mixture proportions. *J. R. Statist. Soc. B* **46**, 465–473.

Hall, P. and Titterington, D.M. (1985). The use of uncategorized data to improve the performance of a nonparametric estimator of a mixture density. *J. R. Statist. Soc. B* **47**, 155–163.

Hall, P. and Welsh, A.H. (1983). A test for normality based on the empirical characteristic function. *Biometrika* **70**, 485–489.

Hampel, F.R. (1973). Robust estimation: a condensed partial survey. *Z. Wahrscheinlickeitstheorie verw. Gebiete* **27**, 87–104.

Hand, D.J. (1981). *Discrimination and Classification.* New York: Wiley.

Hand, D.J. (1982). Kernel Discriminant Analysis. New York: Research Studies Press.

Harding, J.P. (1948). The use of probability paper for the graphical analysis of polymodal frequency distributions. *J. Marine Biol. Assoc. U.K.* **28**, 141–153.

Harris, C.M. (1983). On finite mixtures of geometric and negative binomial distributions. *Commun. Statist.-Theor. Meth.* **12**, 987–1007.

Hartigan, J.A. (1975). *Clustering Algorithms.* New York: Wiley.

Hartigan, J.A. (1977). Distribution problems in clustering. In *Classification and Clustering*, J. Van Ryzin (Ed.). New York: Academic Press, pp. 45–71.

Hartigan, J.A. (1978). Asymptotic distributions for clustering criteria. *Ann. Statist.* **6**. 117–131.

Hartigan, J.A. (1985a). Statistical theory in clustering. *J. Classification*, **2**, 63–76.

Hartigan, J.A. (1985b). A failure of likelihood asymptotics for normal mixtures. Proc. Berkeley Conference in Honor of Jerzy Neyman and Jack Kiefer (Vol. II), L.M. Le Cam and R.A. Olshen (Eds.). Monterey: Wadsworth, pp. 807–810.

Hartigan, J.A. and Hartigan, P.M. (1985). The dip test of unimodality. *Ann. Statist.* **13**, 70–84.

Hartley, H.O. and Rao, J.N.K. (1968). Classification and estimation in analysis of variance problems. *Int. Statist. Rev.* **36**, 141–147.

Hasselblad, V. (1966). Estimation of parameters for a mixture of normal distributions. *Technometrics* **8**, 431–444.

Hasselblad, V. (1969). Estimation of finite mixtures of distributions from the exponential family. *J. Amer. Statist. Assoc.* **64**, 1459–1471.

Hathaway, R.J. (1983). Constrained maximum likelihood estimation for normal mixtures. In *Computer Science and Statistics: The Interface*, J.E. Gentle (Ed.). Amsterdam: North-Holland, pp. 263–267.

Hathaway, R.J. (1985). A constrained formulation of maximum-likelihood estimation for normal mixture distributions. *Ann. Statist.* **13**, 795–800.

Hathaway, R.J. (1986a). Another interpretation of the EM algorithm for mixture distributions. *Statistics & Probability Letters* **4**, 53–56.

Hathaway, R.J. (1986b). A constrained EM algorithm for univariate normal mixtures. *J. Statist. Comput. Simul.* **23**, 211–230.

Hawkins, D.M. (1980). *Identification of Outliers.* London: Chapman and Hall.

Hawkins, D.M. (1981). A new test for multivariate normality and homoscedasticity. *Technometrics* **23**, 105–110.

Hawkins, D.M. (Ed.). (1982). *Topics in Applied Multivariate Analysis.* Cambridge: Cambridge University Press.

Hawkins, D.M., Muller, M.W. and ten Krooden, J.A. (1982). Cluster analysis. In *Topics in Applied Multivariate Analysis*, D.M. Hawkins (Ed.). Cambridge: Cambridge University Press, pp. 303–356.

Heathcote, C.R. (1977). Integrated mean square error estimation of parameters. *Biometrika* **64**, 255–264.

Henna, J. (1985). On estimating of the number of constituents of a finite mixture of continuous distributions. *Ann. Inst. Statist. Math.* **37**, 235–240.

Hernandez-Avila, A. (1979). *Problems in Cluster Analysis.* Unpublished D. Phil. thesis, University of Oxford.

Heydorn, R.P. (1984). Using satellite remotely sensed data to estimate crop proportions. *Commun. Statist.-Theor. Meth.* **13**, 2881–2903.

Hill, B.M. (1963). Information for estimating the proportions in mixtures of exponential and normal distributions. *J. Amer. Statist. Assoc.* **58**, 918–932.

Holgersson, M. and Jorner, U. (1978). Decomposition of a mixture into normal components: a review. *Int. J. Bio-Med. Comput.* **9**, 367–392.

Hope, A.C.A. (1968). A simplified Monte Carlo significance test procedure. *J. R. Statist. Soc. B* **30**, 582–598.

Hosmer, D.W. (1973a). On MLE of the parameters of a mixture of two normal distributions when the sample size is small. *Commun. Statist.* **1**, 217–227.

Hosmer, D.W. (1973b). A comparison of iterative maximum likelihood estimates of the parameters of a mixture of two normal distributions under three different types of sample. *Biometrics* **29**, 761–770.

Hosmer, D.W. (1974). Maximum likelihood estimates of the parameters of a mixture of two regression lines. *Commun. Statist.* **3**, 995–1006.

Hosmer, D.W. (1978). A use of mixtures of two normal distributions in a classification problem. *J. Statist. Comput. Simul.* **6**, 281–294.

Hosmer, D.W. and Dick, N.P. (1977). Information and mixtures of two normal distributions. *J. Statist. Comput. Simul.* **6**. 137–148.

Huber, P.J. (1964). Robust estimation of a location parameter. *Ann. Math. Statist.* **35**, 73–101.

Huber, P.J. (1977). Robust covariances. In *Statistical Decision Theory and Related Topics* (Vol. II), S.S. Gupta and D.S. Moore (Eds.). New York: Academic Press, pp. 165–191.

Huber, P.J. (1981). *Robust Statistics.* New York: Wiley.

Huzurbazar, V.S. (1948). The likelihood equation, consistency, and the maxima of the likelihood function. *Ann. Eugenics* **14**, 185–200.

James, I.R. (1978). Estimation of the mixing proportion in a mixture of two normal distributions from simple, rapid measurements. *Biometrics* **34**, 265–275.

Jardine, N. and Sibson, R. (1971). *Numerical Taxonomy.* London: Wiley.

Jeffreys, H. (1932). An alternative to the rejection of observations. *Proc. Roy. Soc. London A* **137**, 78–87.

John, S. (1970). On identifying the population of origin of each observation in a mixture of observations from two normal populations. *Technometrics* **12**, 553–563.

Johnson, N.L. (1973). Some simple tests of mixtures with symmetrical components. *Commun. Statist.* **1**, 17–25.

Jolliffe, I.T. (1975). Cluster analysis as a multiple comparison method. In *Applied Statistics*, R.P. Gupta (Ed.). Amsterdam: North-Holland, pp. 159–168.

Kanji, G.K. (1985). A mixture model for wind shear data. *J. Appl. Statist.* **12**, 49–58.

Kendall, M.G. (1965). *A Course in Multivariate Analysis.* London: Charles Griffin.

Keuls, M. (1952). The use of the Studentized range in connection with an analysis of variance. *Euphytica* **1**, 112–122.

Kiefer, J. and Wolfowitz, J. (1956). Consistency of the maximum likelihood estimates in the presence of infinitely many incidental parameters. *Ann. Math. Statist.* **27**, 887–906.

Kiefer, N.M. (1978). Discrete parameter variation: efficient estimation of a switching regression model. *Econometrica* **46**, 427–434.

Kittler, J. and Pau, L.F. (1978). Small sample properties of a pattern recognition system in lot acceptance sampling. *Proc.* $4^{\text{th}}$ *Int. Joint Conference on Pattern Recognition, Kyoto.* New York: IEEE, pp. 249–257.

Koziol, J.A. (1982). A class of invariant procedures for assessing multivariate normality. *Biometrika* **69**, 423–427.

Koziol, J.A. (1983). On assessing multivariate normality, *J. R. Statist. Soc. B* **45**, 358–361.

Koziol, J.A. (1986). Assessing multivariate normality: a compendium. *Commun. Statist.-Theor. Meth.* **15**, 2763–2783.

Krishnaiah, P.R. and Kanal, L.N. (Eds.). (1982). *Classification, Pattern Recognition and Reduction of Dimensionality. Handbook of Statistics (Vol. 2).* Amsterdam: North-Holland.

Kruskal, J.B. (1964a). Multidimensional scaling by optimizing goodness of fit to a nonmetric hypothesis. *Psychometrika* **29**, 1–27.

Kruskal, J.B. (1964b). Nonmetric multidimensional scaling: a numerical method. *Psychometrika* **29**, 115–129.

Kruskal, J.B. (1977). The relationship between multidimensional scaling and clustering. In *Classification and Clustering*, J. Van Ryzin (Ed.). New York: Academic Press, pp. 17–44.

Kruskal, J.B. and Wish, M. (1978). *Multidimensional Scaling.* Sage University Paper Series on Quantitative Applications in the Social Sciences. Beverly Hills: Sage Publications.

Kshirsagar, A.M. (1972). *Multivariate Analysis.* New York: Marcel Dekker.

Kuiper, F.K. and Fisher, L. (1975). A Monte Carlo comparison of six clustering procedures. *Biometrics* **31**, 777–783.

Kumar, K.D., Nicklin, E.H. and Paulson, A.S. (1979). Comment on *Estimating Mixtures of Normal Distributions and Switching Regressions*. *J. Amer. Statist. Assoc.* **74**, 52–55.

Lachenbruch, P.A. (1975). *Discriminant Analysis.* New York: Hafner Press.

Lachenbruch, P.A. and Mickey, M.R. (1968). Estimation of error rates in discriminant analysis. *Technometrics* **10**, 1–11.

Laird, N. (1978). Nonparametric maximum likelihood estimation of a mixing distribution. *J. Amer. Statist. Assoc.* **73**, 805–811.

Larson, M.G. and Dinse, G.E. (1985). A mixture model for the regression analysis of competing risks data. *Appl. Statist.* **34**, 201–211.

Lazarsfeld, P.F. and Henry, N.W. (1968). *Latent Structure Analysis.* New York: Houghton Mifflin.

Lee, K.L. (1979). Multivariate tests for clusters. *J. Amer. Statist. Assoc.* **74**, 708–714.

Lehmann, E.L. (1980). Efficient likelihood estimators. *Amer. Statistician* **34**, 233–235.

Lehmann, E.L. (1983). *Theory of Point Estimation.* New York: Wiley.

Lesaffre, E. (1983). Normality tests and transformations. *Pattern Recognition Letters* **1**, 187–199.

Li, L.A. and Sedransk, N. (1986). Mixtures of distributions: a topological approach (Abstract). *Bull. Inst. Math. Statist.* **15**, 336–337.

Lindsay, B.G. (1983). The geometry of mixture likelihoods: a general theory. *Ann. Statist.* **11**, 86-94.

Lissack, T. and Fu, K.S. (1976). Error estimation in pattern recognition via $L^\alpha$-distance between posterior density functions. *IEEE Trans. Inform. Theory* **IT-22**, 34-45.

Louis, T.A. (1982). Finding the observed information matrix when using the EM algorithm. *J. R. Statist. Soc. B* **44**, 226-233.

Macdonald, P.D.M. (1971). Comment on *An Estimation Procedure for Mixtures of Distributions* by Choi and Bulgren. *J. R. Statist. Soc. B* **33**, 326-329.

Macdonald, P.D.M. (1975). Estimation of finite distribution mixtures. In *Applied Statistics*, R.P.Gupta (Ed.). Amsterdam: North-Holland, pp. 231-245.

Macdonald, P.D.M. and Pitcher, T.J. (1979). Age-groups from size-frequency data: a versatile and efficient method of analyzing distribution mixtures. *J. Fish. Res. Board of Canada* **36**, 987-1001.

Machado, S.G. (1983). Two statistics for testing for multivariate normality. *Biometrika* **70**, 713-718.

MacQueen, J. (1967). Some methods for classification and analysis of multivariate observations. *Proc. 5$^{th}$ Berkeley Symp. (Vol. 1).* Berkeley: University of California Press, pp. 281-297.

Manning, H.L. (1956). Yield improvement from a selection index technique with cotton. *Heredity* **10**, 303-322.

Mardia, K.V. (1974). Applications of some measures of multivariate skewness and kurtosis in testing normality and robustness studies. *Sankhyā B* **36**, 115-128.

Mardia, K.V. (1980). Tests of univariate and multivariate normality. In *Handbook of Statistics* (Vol.1), P.R. Krishnaiah (Ed.). Amsterdam: North-Holland, pp. 279-320.

Maronna, R.A. (1976). Robust *M*-estimators of multivariate location and scatter. *Ann. Statist.* **4**, 51–67.

Marriott, F.H.C. (1971). Practical problems in a method of cluster analysis. *Biometrics* **27**, 501–514.

Marriott, F.H.C. (1975). Separating mixtures of normal distributions. *Biometrics* **31**, 767–769.

Marriott, F.H.C. (1982). Optimization methods of cluster analysis. *Biometrika* **69**, 417–421.

Matthews, J.N.S. (1984). Robust methods in the assessment of multivariate normality. *Appl. Statist.* **33**, 272–277.

McLachlan, G.J. (1974). An asymptotic unbiased technique for estimating the error rates in discriminant analysis. *Biometrics* **30**, 239–249.

McLachlan, G.J. (1975a). Iterative reclassification procedure for constructing an asymptotically optimal rule of allocation in discriminant analysis. *J. Amer. Statist. Assoc.* **70**, 365–369.

McLachlan, G.J. (1975b). Confidence intervals for the conditional probability of misallocation in discriminant analysis. *Biometrics* **32**, 161–167.

McLachlan, G.J. (1975c). Some expected values for the error rates of the sample quadratic discriminant function. *Austral. J. Statist.* **17**, 161–165.

McLachlan, G.J. (1976). The bias of the apparent error rate in discriminant analysis. *Biometrika* **63**, 239–244.

McLachlan, G.J. (1977). Estimating the linear discriminant function from initial samples containing a small number of unclassified observations. *J. Amer. Statist. Assoc.* **72**, 403–406.

McLachlan, G.J. (1979). A comparison of the estimative and predictive methods of estimating posterior probabilities. *Commun. Statist.-Theor. Meth.* **A8**, 919–929.

McLachlan, G.J. (1980a). A note on bias correction in maximum likelihood estimation with logistic discrimination. *Technometrics* **22**, 621–627.

McLachlan, G.J. (1980b). The efficiency of Efron's bootstrap approach applied to error rate estimation in discriminant analysis. *J. Statist. Comput. Simul.* **11**, 273–279.

McLachlan, G.J. (1982a). The classification and mixture maximum likelihood approaches to cluster analysis. In *Handbook of Statistics* (Vol. 2), P. R. Krishnaiah and L.N. Kanal (Eds.). Amsterdam: North-Holland, pp. 199–208.

McLachlan, G.J. (1982b). On the bias and variance of some proportion estimators. *Commun. Statist.-Simula. Computa.* **11**, 715–726.

McLachlan, G.J. (1986a). Assessing the performance of an allocation rule. *Comp. & Maths. with Appls.* **12A**, 261–272.

McLachlan, G.J. (1986b). On bootstrapping the likelihood ratio test stastic for the number of components in a normal mixture. *Appl. Statist.* (to appear).

McLachlan, G.J. (1987). Error rate estimation in discriminant analysis: recent advances. In *Advances in Multivariate Statistical Analysis*, A.K. Gupta (Ed.). Dordrecht: Reidel (to appear).

McLachlan, G.J. and Ganesalingam, S. (1982). Updating a discriminant function on the basis of unclassified data. *Commun. Statist-Simula. Computa.* **11**, 753–767.

McLachlan, G.J., Lawoko, C.R.O. and Ganesalingam, S. (1982). On the likelihood ratio test for compound distributions for homogeneity of mixing proportions. *Technometrics* **24**, 331–334.

Menzefricke, U. (1981). Bayesian clustering of data sets. *Commun. Statist.-Theor. Meth.* **A10**, 65–77.

Mezzich, J.E. and Solomon, H. (1980). *Taxonomy and Behavioral Science—Comparative Performance of Grouping Methods.* New York: Academic Press.

Miller, R.G. (1974). The jackknife—a review. *Biometrika* **61**, 1–15.

Miller, R.G. (1981). *Simultaneous Statistical Inference.* Second Edition. New York: Springer-Verlag.

Moran, M.A. and Murphy, B.J. (1979). A closer look at two alternative methods of statistical discrimination. *Appl. Statist.* **28**, 223–232.

Moran, P.A.P. (1973). Asymptotic properties of homogeneity tests. *Biometrika* **60**, 79–85.

Moore, D.S., Whitsitt, S.J. and Landgrebe, D.A. (1976). Variance comparisons for unbiased estimators of probability of correct classification. *IEEE Trans. Inform. Theory* **IT-22**, 102–105.

Morgan, B.J.T. (1981). Three applications of methods of cluster-analysis. *Statistician* **30**, 205–223.

Morrison, D.F. (1976). *Multivariate Statistical Methods.* Second Edition. New York: McGraw-Hill.

Mungomery, V.E., Shorter, R. and Byth, D.E. (1974). Genotype x environment interactions and environmental adaptation. I. Pattern analysis—application to soya bean populations. *Aust. J. Agric. Res.* **25**, 59–72.

Murray, G.D. and Titterington, D.M. (1978). Estimation problems with data from a mixture. *Appl. Statist.* **27**, 325–334.

Nelder, J.A. (1971). Contribution to the discussion of paper by R. O'Neill and G.B. Wetherill. *J. R. Statist. Soc. B* **33**, 244–246.

Newcomb, S. (1886). A generalized theory of the combination of observations so as to obtain the best result. *Amer. J. Math.* **8**, 343–366.

Odell, P.L. (1976). Current and past roles of the statistician in space applications. *Commun. Statist.-Theor. Meth.* **A5**, 1077–1089.

Odell, P.L. and Basu, J.P. (1976). Comparing several methods for estimating crop acreages using remote sensing data. *Commun. Statist.-Theor. Meth.* **A5**, 1091–1114.

O'Neill, R. and Wetherill, G.B. (1971). The present state of multiple comparison methods (with discussion). *J. R. Statist. Soc. B* **33**, 218–250.

O'Neill, T.J. (1976). *Efficiency Calculations in Discriminant Analysis.* Unpublished Ph.D. thesis, Stanford University.

O'Neill, T.J. (1978). Normal discrimination with unclassified observations. *J. Amer. Statist. Assoc.* **73**, 821–826.

Orchard, T. and Woodbury, M.A. (1972). A missing information principle: theory and applications. *Proc.* 6$^{th}$ *Berkeley Symp.* (Vol. 1). Berkeley: University of California Press, pp. 697–715.

Parr, W.C. (1981). Minimum distance estimation: a bibliography. *Commun. Statist.-Theor. Meth.* **A10**, 1205–1224.

Pau, L.F. and Chen, C.H. (1977). Properties of classification rules subject to small learning samples. *Proc.* 1$^{st}$ *Int. Symp. on Data Analysis and Informatics, Versailles.* Rocquencourt-Le Chesnay: IRIA, pp. 532–547.

Pearson, K. (1894). Contributions to the mathematical theory of evolution. *Phil. Trans. A* **185**, 71–110.

Perlman, M.D. (1972). On the strong consistency of approximate maximum likelihood estimators. *Proc.* 6$^{th}$ *Berkeley Symp. (Vol. 1).* Berkeley: University of California Press, pp. 263–281.

Perlman, M.D. (1983). The limiting behavior of multiple roots of the likelihood equation. In *Recent Advances in Statistics*, M.H. Rizvi, J.S. Rustagi and D. Siegmund (Eds.). New York: Academic Press, pp. 339–370.

Perlman, M.D. (1984). Private communication.

Peters, B.C. and Walker, H.F. (1978). An iterative procedure for obtaining maximum-likelihood estimates of the parameters for a mixture of normal distributions. *SIAM J. Appl. Math.* **35**, 362–378.

Plackett, R.L. (1971). Contribution to the discussion of paper by R. O'Neill and G.B. Wetherill. *J. R. Statist. Soc. B* **33**, 242–244.

Quandt, R.E. and Ramsey, J.B. (1978). Estimating mixtures of normal distributions and switching regressions (with discussion). *J. Amer. Statist. Assoc.* **73**, 730–752.

Quenouille, M.H. (1949). Approximate tests of correlation in time-series. *J. R. Statist. Soc. B* **11**, 68–84.

Quenouille, M.H. (1956). Notes on bias in estimation. *Biometrika* **43**, 353–360.

Quinn, B.G., McLachlan, G.J., and Hjort, N.L. (1987). A note on the Aitkin-Rubin approach to hypothesis testing in mixture models. *J. R. Statist. Soc. B* **49** (to appear).

Ramsay, J.O. (1982). Some statistical approaches to multidimensional scaling data (with discussion). *J. R. Statist. Soc. A* **145**, 285–312.

Rao, C.R. (1948). The utilization of multiple measurements in problems of biological classification. *J. R. Statist. Soc. B* **10**, 159–203.

Rao, C.R. (1952). *Advanced Statistical Methods in Biometric Research.* New York: Wiley.

Redner, R.A. (1981). Note on the consistency of the maximum likelihood estimate for nonidentifiable distributions. *Ann. Statist.* **9**, 225–228.

Redner, R.A. and Walker, H.F. (1984). Mixture densities, maximum likelihood and the EM algorithm. *SIAM Review* **26**, 195–239.

Rigby, R.A. (1982). A credibility interval for the probability that a new observation belongs to one of two multivariate normal populations. *J. R. Statist. Soc. B* **44**, 212–220.

Royston, J.P. (1983). Some techniques for assessing multivariate normality based on the Shapiro-Wilk *W*. *Appl. Statist.* **32**, 121–133.

Schmidt, P. (1982). An improved version of the Quandt-Ramsey MGF estimator for mixtures of normal distributions and switching regressions. *Econometrica* **50**, 501–516.

Schwemer, G.T. and Dunn, O.J. (1980). Posterior probability estimators in classification simulations. *Commun. Statist.-Simula. Computa.* **B9**, 133–140.

Sclove, S.L. (1977). Population mixture models and clustering algorithms. *Commun. Statist.-Theor. Meth.* **A6**, 417–434.

Sclove, S.L. (1983). Application of the conditional population-mixture model to image segmentation. *IEEE Trans. Pattern Anal. Machine Intell.* **PAMI-5**, 428–433.

Scott, A.J. and Knott, M. (1974). A cluster analysis method for grouping means in the analysis of variance. *Biometrics* **30**, 507–512.

Scott, A.J. and Symons, M.J. (1971). Clustering methods based on likelihood ratio criteria. *Biometrics* **27**, 387–397.

Seber, G.A.F. (1984). *Multivariate Observations.* New York: Wiley.

Shorter, R., Byth, D.E. and Mungomery, V.E. (1977). Genotype x environment interactions and environmental adaptation. II. Assessment of environmental contributions. *Aust. J. Agric. Res.* **28**, 223–235.

Shepard, R.N. (1962a). The analysis of proximities: multidimensional scaling with an unknown distance function. I. *Psychometrika* **27**, 125–140.

Shepard, R.N. (1962b). The analysis of proximities: multidimensional scaling with an unknown distance function. II. *Psychometrika* **27**, 219–246.

Silverman, B.W. (1981). Using kernel density estimates to investigate multimodality. *J. R. Statist. Soc. B* **43**, 97–99.

Silverman, B.W. (1983). Some properties of a test for multimodality based on kernel density estimates. In *Probability, Statistics and Analysis* (London Math. Soc. Lecture Note Series, 79), J.F.C. Kingman and G.E.H. Reuter (Eds.). Cambridge: Cambridge University Press, pp. 248–259.

Skillings, J.H. (1983). Nonparametric approaches to testing and multiple comparisons in a one-way ANOVA. *Commun. Statist.-Simula. Computa.* **12**, 373–387.

Small, N.J.H. (1980). Marginal skewness and kurtosis in testing multivariate normality. *Appl. Statist.* **29**, 85–87.

Smith, A.F.M. and Makov, U.E. (1978). A quasi-Bayes sequential procedure for mixtures. *J. R. Statist. Soc. B* **40**, 106–112.

Smith, H.F. (1936). A discriminant function for plant selection. *Ann. Eugenics* **7**, 240–250.

Sneath, P.H.A. and Sokal, R.R. (1973). *Numerical Taxonomy. The Principles and Practice of Numerical Classification.* San Francisco: W.H. Freeman.

Srivastava, M.S. (1984). A measure of skewness and kurtosis and a graphical method for assessing multivariate normality. *Statistics & Probability Letters* **2**, 263–267.

Srivastava, M.S. and Lee, G.C. (1984). On the distribution of the correlation coefficient when sampling from a mixture of two bivariate normal densities: robustness and outliers. *Canad. J. Statist.* **2**, 119–133.

Symons, M.J. (1981). Clustering criteria and multivariate normal mixtures. *Biometrics* **37**, 35–43.

Symons, M.J., Grimson, R.C. and Yuan, Y.C. (1983). Clustering of rare events. *Biometrics* **39**, 193–205.

Switzer, P. (1980). Extensions of linear discriminant analysis for statistical classification of remotely sensed imagery. *Math. Geol.* **12**, 367–376.

Switzer, P. (1983). Some spatial statistics for the interpretation of satellite data. *Bull. Int. Statist. Inst.* **50**, 962–971.

Takane, Y., Young, F.W. and de Leeuw, J. (1977). Nonmetric individual differences multidimensional scaling: an alternating least squares method with optimal scaling features. *Psychometrika* **42**, 7–67.

Tarter, M. and Silvers, A. (1975). Implementation and application of bivariate Gaussian mixture decomposition. *J. Amer. Statist. Asssoc.* **70**, 47–55.

Tan, W.Y. and Chang, W.C. (1972). Some comparisons of the method of moments and the method of maximum likelihood in estimating parameters of a mixture of two normal densities. *J. Amer. Statist. Assoc.* **67**, 702–708.

Tibshirani, R. (1985). How many bootstraps? Technical Report No. 362. Stanford: Dept. of Statistics, Stanford University.

Titterington, D.M. (1976). Updating a diagnostic system using unconfirmed cases. *Appl. Statist.* **24**, 238–247.

Titterington, D.M. (1980). A comparative study of kernel-based density estimates for categorical data. *Technometrics* **22**, 259–268.

Titterington, D.M. (1981). Contribution to the discussion of paper by M. Aitkin, D. Anderson and J. Hinde. *J. R. Statist. Soc. A* **144**, 459.

Titterington, D.M. (1983). Minimum distance non-parametric estimation of mixture proportions. *J. R. Statist. Soc. B* **45**, 37–46.

Titterington, D.M. (1984a). Recursive parameter estimation using incomplete data. *J. R. Statist. Soc. B* **46**, 257–267.

Titterington, D.M. (1984b). Comments on *Application of the Conditional Population-Mixture Model to Image Segmentation. IEEE Trans. Pattern Anal. Machine Intell.* **PAMI-6**, 656–658.

Titterington, D.M., Smith, A.F.M. and Makov, U.E. (1985). *Statistical Analysis of Finite Mixture Distributions.* New York: Wiley.

Tubbs, J.D. and Coberly, W.A. (1976). An empirical sensitivity study on mixture proportion estimators. *Commun. Statist.-Theor. Meth.* **A5**, 1115–1125.

Tucker, L.R. and Messick, S. (1963). An individual differences model for multidimensional scaling. *Psychometrika* **28**, 333–367.

Tukey, J.W. (1949). Comparing individual means in the analysis of variance. *Biometrics* **5**, 99–114.

Tukey, J.W. (1958). Bias and confidence in not-quite large samples (Abstract). *Ann. Math. Statist.* **29**, 614.

Tukey, J.W. (1960). A survey of sampling from contaminated distributions. In *Contributions to Probability and Statistics*, I. Olkin, S.G. Ghurye, W. Hoeffding, W.G. Madow and H.B. Mann (Eds.). Stanford: Stanford University Press, pp. 448–485.

Van Ryzin, J. (Ed.). (1977). *Classification and Clustering*. New York: Academic Press.

Wald, A. (1949). Note on the consistency of the maximum likelihood estimate. *Ann. Math. Statist.* **20**, 595–601.

Walker, H.F. (1980). Estimating the proportions of two populations in a mixture using linear maps. *Commun. Statist.-Theor. Meth.* **A9**, 837–849.

White, B.S. and Castleman, K.R. (1981). Estimating cell populations. *Pattern Recognition* **13**, 365–370.

Whitmore, R.C. and Harner, E.J. (1980). Analyses of multivariately determined community matrices using cluster analysis and multidimensional scaling. *Biom. J.* **22**, 715–723.

Wilk, M.B. and Gnanadesikan, R. (1968). Probability plotting methods for the analysis of data. *Biometrika* **55**, 1–17.

Williams, W.T. (1976). Types of classification. In *Pattern Analysis in Agricultural Science*, W.T. Williams (Ed.). Amsterdam: Elsevier, pp. 76–83.

Wilson, E.B. and Hilferty, M.M. (1931). The distribution of chi-square. *Proc. Nat. Acad. Sci. U.S.A.* **28**, 94–100.

Wolfe, J.H. (1967). NORMIX: Computational methods for estimating the parameters of multivariate normal mixtures of distributions. Research Memo. SRM 68-2. San Diego: U.S. Naval Personnel Research Activity.

Wolfe, J.H. (1970). Pattern clustering by multivariate mixture analysis. *Multivar. Behav. Res.* **5**, 329–350.

Wolfe, J.H. (1971). A Monte Carlo study of the sampling distribution of the likelihood ratio for mixtures of multinormal distributions. *Technical Bulletin STB 72-2.* San Diego: U.S. Naval Personnel and Training Research Laboratory.

Wong, M.A. (1985). A bootstrap testing procedure for investigating the number of subpopulations. *J. Statist. Comput. Simul.* **22**, 99–112.

Wong, M.A. and Lane, T. (1983). A $k$th nearest neighbour clustering procedure. *J. R. Statist. Soc. B* **45**, 362–368.

Woodward, W.A., Parr, W.C., Schucany, W.R. and Lindsey, H. (1984). A comparison of minimum distance and maximum likelihood estimation of a mixture proportion. *J. Amer. Statist. Assoc.* **79**, 590–598.

Worlund, D.D. and Fredin, R.A. (1962). Differentiation of stocks. *Symp. on Pink Salmon*, N.J. Wilimovsky (Ed.). Vancouver: Institute of Fisheries, University of British Columbia, pp. 143–153.

Wu, C.F.J. (1983). On the convergence properties of the EM algorithm. *Ann. Statist.* **11**, 95–103.

Young, F.W. and Lewyckyj, R. (1980). *ALSCAL 4 User's Guide.* Chapel Hill: Data Analysis and Theory Associates.

# Appendix

The appendix contains the following FORTRAN programs for fitting a mixture of normal distributions:

(i) KMM fits a mixture of normal distributions with either equal or arbitrary covariance matrices to two-mode two-way data.
(ii) KPCMM fits a mixture of normal distributions with arbitrary covariance matrices to two-mode two-way data in conjunction with at least $p + 1$ classified observations avaliable from each of the possible groups. The program allows for both mixture and separate sampling of the classified data. There is also the provision for the robust fitting of the model.
(iii) KTMM fits a normal mixture model to $n$ treatment sample means having variances $\sigma^2/r_1, \cdots, \sigma^2/r_n$, where $r_1, \ldots, r_n$ are known constants. Available also is an independent estimate of $\sigma^2$.
(iv) K3MM fits a mixture of normal distributions with arbitrary covariance matrices to three-mode three-way data.

Two subroutines GDET and MATINV used in KPCMM are not listed there because they are identical to those listed in KMM. The program K3MM has been converted to use single dimensional arrays so these modified subroutines are included in this listing. The user may wish to substitute another matrix inversion subroutine instead of MATINV. All of the above programs have been run on an IBM mainframe machine.

```
      PROGRAM KMM
C     Program for fitting a mixture of normal distributions with either
C     equal or arbitrary covariance matrices to two-mode two-way data.
C     The first mode is taken to be the class of entities to be
C     clustered.  The second mode is a class of some characteristics
C     of the entities, hereafter referred to as the attributes.
C
C     The input file contains the following:
C     NIND NATT                (number of entities and attributes)
C     (X(I,J),J=1,NATT)
C     ------------------      (NIND rows of data each having
C     ------------------       NATT values per line)
C     ------------------
C     NG                       (number of groups)
C     NCOV                     (1 = Equal covariance matrices;
C                               2 = Arbitrary covariance matrices)
C     NSWH                     (1 = Initially specified partition
C                                   of entities into NG groups;
C                               2 = Initially specified estimates of
C                                   unknown parameters for the NG groups)
C                        If NSWH = 1
C     (IDT(I),I=1,NIND)    (initial grouping of the entities)
C        This allows free field input but for ease of reading use
C        ten numbers per line for I=1,NIND.
C     NOTE:  The initial partition of the entities requires at
C            least NATT+1 entities to be assigned to each group.
C                 or if NSWH = 2
C     (XBAR(K,J),J=1,NATT)    (estimated mean vector for each group)
C     ------------------
C     (XVAR(K,I,J),J=1,NATT) (estimated covariance matrix for each
C     ------------------        group with NATT rows for each matrix)
C     ------------------       If NCOV = 1 there is only one matrix
C     ------------------
C     (T(K),K=1,NG)           (estimated mixing proportion for each group)
C
C     The output file contains the results of fitting the above model
C     starting from an initially specified partition of the data or
C     initially specified estimates of the unknown parameters.
C     It includes the final probabilistic clustering and the
C     consequent partition of the NIND entities into NG groups.
C
C     This section of the program calls the subroutines and then
C     writes out some summary results.
      DIMENSION X(200,10),XVAR(10,10,10),T(10),DV(10),V(10,10,10),
     1XMU(10,10)
      CALL GIN(NIND,NATT,NG,X,XMU,V,DV,T,NCOV,IER)
      IF (IER.NE.0) GO TO 605
      CALL LOOP(NIND,NATT,NG,X,XMU,V,XVAR,DV,T,NCOV,IER)
      IF (IER.NE.0) GO TO 605
      WRITE (22,106)
106   FORMAT (//2X,'Estimated mean (as a row vector) for each group')
      DO 20 K=1,NG
20    WRITE (22,103) (XMU(K,J),J=1,NATT)
103   FORMAT (2X,10F13.6)
C     Test if a common covariance matrix is specified (NCOV = 1)
      IF (NCOV.EQ.1) GO TO 600
      DO 30 K=1,NG
      WRITE (22,105) K
105   FORMAT (//2X,'Estimated covariance matrix for group ',I2)
      DO 30 J=1,NATT
30    WRITE (22,104) (XVAR(K,I,J),I=1,J)
104   FORMAT (5X,10F13.6)
      GO TO 610
600   WRITE (22,107)
```

```
107     FORMAT (//2X,'Estimated common covariance matrix ')
        DO 40 J=1,NATT
40      WRITE (22,104) (XVAR(1,I,J),I=1,J)
        GO TO 610
605     WRITE (22,101) IER
101     FORMAT (//2X,'Terminal error in MATINV as IERR = ',I6)
610     STOP
        END


        SUBROUTINE GIN(NIND,NATT,NG,X,XBAR,V,DV,T,NCOV,IER)
C       This subroutine reads the data and depending on the value of NSWH
C       either (i) calculates estimates of the mean, covariance matrix
C                  and mixing proportion for each group, corresponding
C                  to the initial partition of the NIND entities into
C                  NG groups
C       or    (ii) reads initial estimates of the unknown parameters.
        DIMENSION X(200,10),XBAR(10,10),XVAR(10,10,10),IDT(200),N(10),
       1T(10),V(10,10,10),AL(10),DV(10),XV(10,10),XSUM(10,10)
        READ (21,*) NIND,NATT
        DO 10 I=1,NIND
10      READ (21,*) (X(I,J),J=1,NATT)
        READ (21,*) NG
        READ (21,*) NCOV
        READ (21,*) NSWH
        IF (NSWH.EQ.2) GO TO 200
C       This section is appropriate for NSWH = 1
        READ (21,*) (IDT(I),I=1,NIND)
        WRITE (22,109)
109     FORMAT (//2X,'Initial partition as specified by input')
        WRITE (22,110) (IDT(I),I=1,NIND)
110     FORMAT (2X,10I4)
C       Compute estimates of mean, covariance matrix and mixing
C       proportion for each group
71      DO 15 K=1,NG
        N(K)=0
        DO 15 J=1,NATT
        XBAR(K,J)=0.0
        DO 15 I=1,NATT
15      XVAR(K,I,J)=0.0
        WRITE (22,106)
106     FORMAT (//2X,'Estimated mean (as a row vector) for each group')
        DO 20 JJ=1,NIND
        K=IDT(JJ)
        N(K)=N(K)+1
        DO 20 J=1,NATT
        XBAR(K,J)=XBAR(K,J)+X(JJ,J)
        DO 20 I=1,J
20      XVAR(K,I,J)=XVAR(K,I,J)+X(JJ,J)*X(JJ,I)
        DO 35 K=1,NG
        DO 30 J=1,NATT
30      XBAR(K,J)=XBAR(K,J)/N(K)
        T(K)=N(K)
        T(K)=T(K)/FLOAT(NIND)
35      WRITE (22,103) (XBAR(K,J),J=1,NATT)
103     FORMAT (2X,10F13.6)
C       Compute estimate of covariance matrix for each group
        DO 60 K=1,NG
        DO 50 J=1,NATT
        DO 50 I=1,J
```

```
      XVAR(K,I,J)=XVAR(K,I,J)-N(K)*XBAR(K,J)*XBAR(K,I)
      XVAR(K,I,J)=XVAR(K,I,J)/FLOAT(N(K)-1)
50    XVAR(K,J,I)=XVAR(K,I,J)
      WRITE (22,105) K
105   FORMAT (//2X,'Estimated covariance matrix for group ',I4)
      DO 60 J=1,NATT
60    WRITE (22,103) (XVAR(K,I,J),I=1,J)
C     Test if a common covariance matrix is specified (NCOV = 1)
      IF (NCOV.EQ.1) GO TO 600
C     Obtain inverse and determinant of each estimated covariance
C     matrix
      CALL GDET(NATT,NG,XVAR,V,DV,IER)
      IF (IER.NE.0) RETURN
      GO TO 412
600   CONTINUE
C     Compute pooled estimate of common covariance matrix
      DO 875 J=1,NATT
      DO 875 I=1,J
      XV(I,J)=0.0
      DO 874 K=1,NG
874   XV(I,J)=XV(I,J)+(N(K)-1)*XVAR(K,I,J)
      XVAR(1,I,J)=XV(I,J)/FLOAT(NIND-NG)
875   XVAR(1,J,I)=XVAR(1,I,J)
      WRITE (22,115)
115   FORMAT (//2X,'Estimated common covariance matrix ')
      DO 878 J=1,NATT
878   WRITE (22,103) (XVAR(1,I,J),I=1,J)
C     Obtain inverse of this matrix
      CALL GDET(NATT,1,XVAR,V,DV,IER)
      IF (IER.NE.0) RETURN
      DO 411 K=2,NG
      DV(K)=DV(1)
      DO 411 J=1,NATT
      DO 411 I=1,NATT
      XVAR(K,I,J)=XVAR(1,I,J)
411   V(K,I,J)=V(1,I,J)
412   CONTINUE
      WRITE (22,112)
112   FORMAT (//2X,'Proportion from each group as specified by input')
      WRITE (22,111) (T(K),K=1,NG)
111   FORMAT (5X,10F7.3)
      RETURN
200   CONTINUE
C     This section is appropriate for NSWH = 2
      WRITE (22,106)
206   FORMAT (//2X,'Estimated mean (as a row vector) for each group')
      DO 80 K=1,NG
      READ (21,*) (XBAR(K,J),J=1,NATT)
80    WRITE (22,103) (XBAR(K,J),J=1,NATT)
C     Test if a common covariance matrix is specified (NCOV = 1)
      IF (NCOV.EQ.1) GO TO 700
      DO 90 K=1,NG
      WRITE (22,107) K
107   FORMAT (//2X,'Estimated covariance matrix for group ',I2)
      DO 90 J=1,NATT
      READ (21,*) (XVAR(K,I,J),I=1,J)
      DO 85 I=1,J
85    XVAR(K,J,I)=XVAR(K,I,J)
90    WRITE (22,103) (XVAR(K,I,J),I=1,J)
      CALL GDET(NATT,NG,XVAR,V,DV,IER)
```

```
      IF (IER.NE.0) RETURN
      GO TO 812
700   CONTINUE
C     Read estimated common covariance matrix
      WRITE (22,113)
113   FORMAT (//2X,'Estimated common covariance matrix ')
      DO 95 J=1,NATT
      READ (21,*) (XVAR(1,I,J),I=1,J)
      DO 93 I=1,J
93    XVAR(1,J,I)=XVAR(1,I,J)
95    WRITE (22,103) (XVAR(1,I,J),I=1,J)
      CALL GDET(NATT,1,XVAR,V,DV,IER)
      IF (IER.NE.0) RETURN
      DO 96 K=2,NG
      DV(K)=DV(1)
      DO 96 J=1,NATT
      DO 96 I=1,NATT
      XVAR(K,I,J)=XVAR(1,I,J)
96    V(K,I,J)=V(1,I,J)
812   CONTINUE
C     Read estimated mixing proportion for each group
      READ (21,*) (T(K),K=1,NG)
      WRITE (22,108)
108   FORMAT (//2X,'Estimated mixing proportion for each group')
      WRITE (22,111) (T(K),K=1,NG)
      RETURN
      END
      SUBROUTINE GDET(NATT,NG,XVAR,V,DV,IER)
C     This subroutine reads all covariance matrices, then calls
C     MATINV, which inverts a matrix and calculates its determinant,
C     for each covariance matrix in turn.
      DIMENSION XVAR(10,10,10),V(10,10,10),DV(10),E(10,10),IP(10)
      DO 40 K=1,NG
      DO 20 J=1,NATT
      DO 20 I=1,J
      E(I,J)=XVAR(K,I,J)
20    E(J,I)=E(I,J)
      TOL=0.000001
      CALL MATINV(E,NATT,NATT,10,DET,IP,TOL,IER)
      IF (IER.NE.0) RETURN
      DO 35 J=1,NATT
      DO 35 I=1,J
      V(K,I,J)=E(I,J)
      V(K,J,I)=V(K,I,J)
35    CONTINUE
      DV(K)=DET
40    CONTINUE
      END


      SUBROUTINE LOOP(NIND,NATT,NG,X,XMU,V,XVAR,DV,T,NCOV,IER)
C     This subroutine uses the EM algorithm from a specified starting
C     value to find a solution of the likelihood equation.
      DIMENSION X(200,10),V(10,10,10),AL(10),DV(10),T(10),N(10),
     1W(200,10),WSUM(10),SUM(10,10),V1(10,10,10),XVAR(10,10,10),
     2IDT(200),XLOGL(150),XCC(10),XMU(10,10),XV(10,10)
      PI=3.141592653
      IOUNT=1
1111  CONTINUE
C     Compute the log likelihood
```

```
      XLOGL(IOUNT)=0.0
      DO 205 JJ=1,NIND
      GUM=0.0
      DO 800 K=1,NG
      AL(K)=0.0
      DO 805 I=1,NATT
      DO 805 J=1,NATT
      AL(K)=AL(K)+(X(JJ,I)-XMU(K,I))*V(K,I,J)*(X(JJ,J)-XMU(K,J))
805   CONTINUE
      IF (AL(K).GT.75.0) GO TO 8055
      AL(K)=-0.5*AL(K)
      AL(K)=EXP(AL(K))/(SQRT(DV(K))*(2.*PI)**(NATT/2.0))
      GO TO 898
8055  AL(K)=0.0
898   CONTINUE
      C=1.0E-30
      IF (T(K).LT.C.OR.AL(K).LT.C) GO TO 800
      GUM=GUM+T(K)*AL(K)
800   CONTINUE
C     Compute current estimates of posterior probabilities of group
C     membership (W)
      IF (GUM.EQ.0.0) GO TO 8110
      DO 810 K=1,NG
      IF (T(K).LT.C.OR.AL(K).LT.C) GO TO 901
      W(JJ,K)=T(K)*AL(K)/GUM
      GO TO 810
901   W(JJ,K)=0.0
810   CONTINUE
      XLOGL(IOUNT)=XLOGL(IOUNT)+ALOG(GUM)
      GO TO 205
8110  DO 8101 K=1,NG
8101  W(JJ,K)=0.0
      WRITE(22,116) IOUNT,JJ
116   FORMAT (/2X,'In loop ',I3,' the estimated mixture density is ',
     1'zero ',/4X,'for the observation on entity ',I6)
205   CONTINUE
C     Test for exit from loop
      IF (IOUNT.LE.10) GO TO 230
      LAST=IOUNT-10
      ALIM=0.0001*XLOGL(LAST)
      DIFF=XLOGL(IOUNT)-XLOGL(LAST)
      IF (ABS(DIFF).LE.ABS(ALIM)) GO TO 990
      IF (IOUNT.GT.149) GO TO 985
230   CONTINUE
C     Compute new estimate of mixing proportion (T) for each group
      DO 666 K=1,NG
      WSUM(K)=0.0
      DO 667 JJ=1,NIND
      WSUM(K)=WSUM(K)+W(JJ,K)
667   CONTINUE
      T(K)=WSUM(K)/NIND
666   CONTINUE
C     Compute new estimates of group means (XMU)
      DO 671 K=1,NG
      DO 671 J=1,NATT
      SUM(K,J)=0.0
      DO 670 JJ=1,NIND
      SUM(K,J)=SUM(K,J)+X(JJ,J)*W(JJ,K)
670   CONTINUE
      XMU(K,J)=SUM(K,J)/WSUM(K)
```

```
671     CONTINUE
C       Compute new estimate of covariance matrix for each group
        DO 870 K=1,NG
        DO 871 J=1,NATT
        DO 871 I=1,J
        V(K,I,J)=0.0
871     CONTINUE
        DO 872 JJ=1,NIND
        DO 873 J=1,NATT
        DO 873 I=1,J
        V(K,I,J)=V(K,I,J)+(X(JJ,I)-XMU(K,I))*(X(JJ,J)-XMU(K,J))*W(JJ,K)
873     CONTINUE
872     CONTINUE
        DO 673 J=1,NATT
        DO 673 I=1,J
        V(K,I,J)=V(K,I,J)/WSUM(K)
        V(K,J,I)=V(K,I,J)
673     CONTINUE
870     CONTINUE
C       Test if a common covariance matrix is specified (NCOV = 1)
        IF (NCOV.EQ.1) GO TO 600
C       Obtain inverse and determinant of each estimated covariance
C       matrix
        CALL GDET(NATT,NG,V,V1,DV,IER)
        IF (IER.NE.0) RETURN
        DO 410 K=1,NG
        DO 410 J=1,NATT
        DO 410 I=1,NATT
        XVAR(K,J,I)=V(K,J,I)
410     V(K,J,I)=V1(K,J,I)
        GO TO 412
600     CONTINUE
C       Compute new estimate of common covariance matrix
        DO 875 J=1,NATT
        DO 875 I=1,J
        XV(I,J)=0.0
        DO 874 K=1,NG
874     XV(I,J)=XV(I,J)+V(K,I,J)*WSUM(K)
        V(1,I,J)=XV(I,J)/NIND
875     V(1,J,I)=V(1,I,J)
C       Obtain inverse of this matrix
        CALL GDET(NATT,1,V,V1,DV,IER)
        IF (IER.NE.0) RETURN
        DO 411 K=1,NG
        DV(K)=DV(1)
        DO 411 J=1,NATT
        DO 411 I=1,NATT
411     XVAR(K,I,J)=V(1,I,J)
        DO 413 K=1,NG
        DO 413 J=1,NATT
        DO 413 I=1,NATT
413     V(K,I,J)=V1(1,I,J)
412     CONTINUE
        IOUNT=IOUNT+1
        GO TO 1111
985     CONTINUE
        WRITE (22,115)
115     FORMAT (//2X,'Note: This sample did not converge in 150 ',
      1'iterations.',/8X,'However the program will continue to ',
      2'print results ',/8X,'obtained from the last cycle estimates.')
```

```
990   CONTINUE
      WRITE (22,611) IOUNT,XLOGL(IOUNT)
611   FORMAT (//2X,'In loop ',I3,' log likelihood is ',F15.3)
      WRITE (22,612)
612   FORMAT (/4X,'Estimate of mixing proportion for each group')
      WRITE (22,102) (T(K),K=1,NG)
102   FORMAT (2X,10F7.3)
C     Determine partition of entities into NG groups
      DO 30 K=1,NG
30    XCC(K)=0.0
      DO 50 I=1,NIND
      MAX=1
      DO 40 K=2,NG
40    IF (W(I,K).GT.W(I,MAX)) MAX=K
      XCC(MAX)=XCC(MAX)+W(I,MAX)
      IDT(I)=MAX
50    CONTINUE
      WRITE (22,101)
101   FORMAT (//2X,'Entity: Final estimates of posterior ',
     1'probabilities of group membership'/)
      DO 60 I=1,NIND
60    WRITE (22,103) I,(W(I,K),K=1,NG)
103   FORMAT (2X,I6,2X,10F7.3)
      WRITE (22,107)
107   FORMAT (//2X,'Resulting partition of the entities into NG groups')
      WRITE (22,108) (IDT(I),I=1,NIND)
108   FORMAT (2X,10I4)
      DO 310 K=1,NG
310   N(K)=0
      DO 320 I=1,NIND
      K=IDT(I)
      IF (K.EQ.0) GO TO 320
      N(K)=N(K)+1
320   CONTINUE
      WRITE (22,113)
113   FORMAT (//2X,'Number assigned to each group')
      WRITE (22,114) (N(K),K=1,NG)
114   FORMAT (2X,10I6)
C     Compute estimates of correct allocation rates
      CC=0.0
      DO 70 K=1,NG
      XCC(K)=XCC(K)/(NIND*T(K))
70    CC=CC+T(K)*XCC(K)
      WRITE (22,105)
105   FORMAT (//2X,'Estimates of correct allocation rates for ',
     1'each group')
      WRITE (22,102) (XCC(K),K=1,NG)
      WRITE (22,104) CC
104   FORMAT (//2X,'Estimate of overall correct allocation rate ',
     1F7.3)
      RETURN
      END


      SUBROUTINE MATINV(A,NR,NC,NMAX,DET,IP,TOL,IERR)
      DIMENSION A(10,11),IP(10)
C     This is a matrix inversion subroutine originally written by
C     I. Oliver of the University of Queensland Computer Centre in
C     September 1966 in FORTRAN IV.  It was revised by J. Williams in
```

```
C      April 1967; then revised and updated by I. Oliver in November
C      1969.  Ref. UQ D4.505
C
C      Note:   NMAX on input must equal the row dimension of A
C              In this program it is set at 10.
C
C      Initialization
       IERR=0
       IF (NR.LE.NMAX.AND.NR.GT.0.AND.NC.GT.0) GO TO 5
       IERR=2
       RETURN
5      DET=1.0
       DO 10 I=1,NR
10     IP(I)=I
       DO 170 K=1,NR
C      Search for pivot
       IR=K
       AMAX=0.0
       DO 20 I=K,NR
       IF (ABS(A(I,K)).LE.ABS(AMAX)) GO TO 20
       AMAX=A(I,K)
       IR=I
20     CONTINUE
       DET=DET*AMAX
C      Record pivot row
       I=IP(K)
       IP(K)=IP(IR)
       IP(IR)=I
C      Test for zero pivot
       IF (AMAX.NE.0.) GO TO 70
       IERR=1
       RETURN
C      Move selected pivot row to pivot row
70     IF (IR.EQ.K) GO TO 100
       DO 90 J=1,NC
       TEMP=A(IR,J)
       A(IR,J)=A(K,J)
90     A(K,J)=TEMP
       DET=-DET
C      Transform matrix
100    DO 110 J-1,NC
110    A(K,J)=A(K,J)/AMAX
       DO 150 I=1,NR
       IF (I.EQ.K) GO TO 150
       DO 140 J=1,NC
       IF (J.EQ.K) GO TO 140
       T1=A(I,J)
       T2=A(I,K)*A(K,J)
       T3-T1-T2
       IF (ABS(T3).LT.(ABS(T1)+ABS(T2))*TOL)T3=0.0
       A(I,J)=T3
140    CONTINUE
150    CONTINUE
       DO 160 I=1,NR
160    A(I,K)=-A(I,K)/AMAX
       A(K,K)=1.0/AMAX
170    CONTINUE
C      Realign inverse
       DO 220 K=1,NR
       DO 180 J=1,NR
       IF (K.EQ.IP(J)) GO TO 190
180    CONTINUE
```

```
190   IF (J.EQ.K) GO TO 220
      DO 210 I=1,NR
      TEMP=A(I,J)
      A(I,J)=A(I,K)
210   A(I,K)=TEMP
      I=IP(K)
      IP(K)=IP(J)
      IP(J)=I
220   CONTINUE
      IF (IERR.EQ.1) IERR=3
      IF (IERR.EQ.2) IERR=0
      RETURN
      END


      PROGRAM KPCMM
C     Program for fitting a mixture of normal distributions with
C     arbitrary covariance matrices to two-mode two-way unclassified
C     data (X) in conjunction with at least NATT+1 classified
C     entities from each of the possible groups.
C     There is provision for the robust fitting of the model.
C     The classified data (Y) may have been obtained either by
C     sampling from the same mixture from which the unclassified data
C     were obtained (mixture sampling, NSAM=1) or by a sampling scheme
C     whereby no information is provided on the mixing proportions
C     (NSAM=2), for example, by sampling separately from each of the
C     groups.  This case NSAM=2 is referred to as separate sampling,
C     although obviously it might arise in other ways, for example,
C     where the classified data were obtained by sampling from a
C     mixture with different mixing proportions to those of the mixture
C     from which the unclassified data were sampled.  The program uses
C     the classified data to produce initial estimates of the group
C     means and covariance matrices, and also of the mixing proportions
C     in the case of mixture sampling.  For the separate sampling case,
C     the mixing proportions are initially taken to be equal to each
C     other.
C
C     The input file contains the following:
C     MIND NATT             (number of classified entities and attributes)
C     (Y(I,J),J=1,NATT)
C     ------------------    (MIND rows of classified data  Y
C     ------------------     each having NATT values per line)
C     ------------------
C     NG                    (number of groups)
C     (IDTT(I),I=1,MIND)    (group identification of classified entities)
C     NSAM                  (1 = Mixture sampling; 2 = Separate sampling)
C     NIND                  (number of unclassified entities)
C     (X(I,J),J=1,NATT)
C     ------------------    (NIND rows of unclassified data  X
C     ------------------     each having NATT values per line)
C     ------------------
C     NROB                  (1 = Robust solution; 2 = Normal solution)
C
C     The output file contains the results of fitting the above model
C     to the unclassified data in conjunction with the classified
C     data.  It includes the final probabilistic clustering and the
C     consequent partition of the NIND entities into the NG groups as
C     well as separately listing those entities not allocated to any
C     existing group.
C
C     This section of the program calls the subroutines and then
C     writes out some summary results.
```

```
      DIMENSION Y(200,10),YVAR(10,10,10),T(10),M(10),DV(10),
     1V(10,10,10),IDTT(200),XVAR(10,10,10),XCC(10),YMU(10,10),
     2X(1200,10),IDT(1200)
      CALL GIN(MIND,NATT,NG,Y,YMU,YVAR,V,DV,IER,M,T,IDTT,NIND,X,NSAM
     1NROB)
      IF (IER.NE.0) GO TO 605
      CALL LOOP(MIND,NATT,NG,Y,YMU,YVAR,V,XVAR,DV,M,T,NIND,X,IER,
     1IDTT,NSAM,NROB)
      IF (IER.NE.0) GO TO 605
      WRITE (22,106)
106   FORMAT (//2X,'Estimated mean (as a row vector) for each group')
      DO 20 K=1,NG
20    WRITE (22,103) (YMU(K,J),J=1,NATT)
103   FORMAT (2X,10F13.6)
      DO 30 K=1,NG
      WRITE (22,105) K
105   FORMAT (//2X,'Estimated covariance matrix for group ',I2)
      DO 30 J=1,NATT
30    WRITE (22,104) (XVAR(K,I,J),I=1,J)
104   FORMAT (5X,10F13.6)
      GO TO 610
605   WRITE (22,101) IER
101   FORMAT (//2X,'Terminal error in MATINV as IERR = ',I6)
610   STOP
      END


      SUBROUTINE GIN(MIND,NATT,NG,Y,YBAR,YVAR,V,DV,IER,M,T,IDTT,NIND,
     1X,NSAM,NROB)
C     This subroutine reads the data and calculates estimates of the
C     mean, covariance matrix and mixing proportion for each group,
C     corresponding to the known origin of the MIND entities from NG
C     groups.  It then reads the data on the NIND unclassified
C     entities.
      DIMENSION Y(200,10),YBAR(10,10),YVAR(10,10,10),IDT(1200),M(10),
     1T(10),IDTT(200),X(1200,10),V(10,10,10),DV(10)
      READ (21,*) MIND,NATT
      DO 10 I=1,MIND
10    READ (21,*) (Y(I,J),J=1,NATT)
      READ (21,*) NG
      READ (21,*) (IDTT(I),I=1,MIND)
      WRITE (22,109)
109   FORMAT (//2X,'Individual group origin of classified data')
      WRITE (22,110) (IDTT(I),I=1,MIND)
110   FORMAT (2X,10I4)
      READ (21,*) NSAM
C     Compute estimates of mean, covariance matrix and, where
C     appropriate, mixing proportion for each group
71    DO 15 K=1,NG
      M(K)=0
      DO 15 J=1,NATT
      YBAR(K,J)=0.0
      DO 15 I=1,NATT
15    YVAR(K,I,J)=0.0
      WRITE (22,106)
106   FORMAT (//2X,'Estimated mean (as a row vector) for each group')
      DO 20 JJ=1,MIND
      K=IDTT(JJ)
      M(K)=M(K)+1
      DO 20 J=1,NATT
      YBAR(K,J)=YBAR(K,J)+Y(JJ,J)
      DO 20 I=1,J
```

```
20      YVAR(K,I,J)=YVAR(K,I,J)+Y(JJ,J)*Y(JJ,I)
        DO 35 K=1,NG
        DO 30 J=1,NATT
30      YBAR(K,J)=YBAR(K,J)/M(K)
        T(K)=M(K)
        T(K)=T(K)/FLOAT(MIND)
35      WRITE (22,103) (YBAR(K,J),J=1,NATT)
103     FORMAT (2X,10F13.6)
        DO 60 K=1,NG
        DO 50 J=1,NATT
        DO 50 I=1,J
        YVAR(K,I,J)=YVAR(K,I,J)-M(K)*YBAR(K,J)*YBAR(K,I)
        YVAR(K,I,J)=YVAR(K,I,J)/FLOAT(M(K)-1)
50      YVAR(K,J,I)=YVAR(K,I,J)
        WRITE (22,105) K
105     FORMAT (//2X,'Estimated covariance matrix for group ',I4)
        DO 60 J=1,NATT
60      WRITE (22,103) (YVAR(K,I,J),I=1,J)
C       Obtain inverse and determinant of each estimated covariance
C       matrix
        CALL GDET(NATT,NG,YVAR,V,DV,IER)
        IF (IER.NE.0) RETURN
        WRITE (22,108)
108     FORMAT (//2X,'Number of classified data from each group')
        WRITE (22,107) (M(K),K=1,NG)
107     FORMAT (5X,10I6)
        WRITE (22,112)
112     FORMAT (//2X,'Proportion of classified data from each group')
        WRITE (22,111) (T(K),K=1,NG)
111     FORMAT (5X,10F7.3)
        IF (NSAM.EQ.1) GO TO 65
        WRITE (22,113)
113     FORMAT (//2X,'Classified data obtained by separate sampling')
        DO 55 K=1,NG
55      T(K)=1/FLOAT(NG)
        WRITE (22,115)
115     FORMAT (//2X,'Initial estimates of mixing proportions')
        WRITE (22,111) (T(K),K=1,NG)
        GO TO 68
65      WRITE (22,114)
114     FORMAT (//2X,'Classified data obtained by mixture sampling')
68      CONTINUE
        READ (21,*) NIND
        DO 70 I=1,NIND
70      READ (21,*) (X(I,J),J=1,NATT)
        READ (21,*) NROB
        IF (NROB.EQ.1) GO TO 75
        WRITE (22,116)
116     FORMAT (//2X,'Normal solution required')
        GO TO 80
75      WRITE (22,117)
117     FORMAT (//2X,'Robust solution required')
80      CONTINUE
        RETURN
        END


        SUBROUTINE LOOP(MIND,NATT,NG,Y,YMU,YVAR,V,XVAR,DV,M,T,NIND,X,
       1IER,IDTT,NSAM,NROB)
C       This subroutine proceeds iteratively from a specified starting
C       value, corresponding to an estimate based solely on the
C       classified data, to find a solution (robust if required) of
```

```
C       the likelihood equation.
        DIMENSION Y(200,10),V(10,10,10),AL(10),DV(10),T(10),M(10),
       1W(1200,10),WSUM(10),SUM(10,10),V1(10,10,10),XVAR(10,10,10),
       2N(10),IDT(1200),XLOGL(150),XCC(10),Z(200,10),IDTT(200),
       3X(1200,10),YMU(10,10),YVAR(10,10,10),IOL(200),
       4WSUMM(10),XSUM(10),TAU(10,1200),WW(10,1200),
       5WY(10,200),YSUM(10),YYSUM(10),SUMY(10,10),VY(10,10,10)
        PI=3.141592653
C       Calculate tuning constant to be used in robust solution
        TIP=2.0/(9.0*NATT)
        PP=SQRT(NATT*(1.0+SQRT(TIP)*1.645-TIP)**3)
        IOUNT=1
1111    CONTINUE
C       Compute the log likelihood under the normal mixture model
        XLOGL(IOUNT)=0.0
C       For each classified entity, compute the Mahalanobis distance
C       between it and the estimated mean of the group to which it
C       belongs
        DO 206 JJ=1,MIND
        DO 206 K=1,NG
        AL(K)=0.0
        IF (K.NE.IDTT(JJ)) GO TO 807
        DO 806 I=1,NATT
        DO 806 J=1,NATT
        AL(K)=AL(K)+(Y(JJ,I)-YMU(K,I))*V(K,I,J)*(Y(JJ,J)-YMU(K,J))
806     CONTINUE
C       Apply Huber's weight function if robust solution required
        IF (NROB.NE.1) GO TO 904
        TAU(K,JJ)=SQRT(AL(K))
        IF (TAU(K,JJ).LT.PP) WY(K,JJ)=1.0
        IF (TAU(K,JJ).GE.PP) WY(K,JJ)=PP/TAU(K,JJ)
        GO TO 905
904     WY(K,JJ)=1.0
905     CONTINUE
        IF (AL(K).GT.75.0) GO TO 8155
        AL(K)=-0.5*AL(K)
        AL(K)=EXP(AL(K))/(SQRT(DV(K))*(2.*PI)**(NATT/2.0))
        GO TO 8988
8155    AL(K)=0.0
8988    CONTINUE
        C=1.0E-30
        IF (AL(K).LT.C) GO TO 206
        XLOGL(IOUNT)=XLOGL(IOUNT)+ALOG(AL(K))
        GO TO 206
807     WY(K,JJ)=0.0
206     CONTINUE
C       Under mixture sampling (NSAM=1), the classified data contribute
C       to the log likelihood also with respect to the mixing proportions
        IF (NSAM.NE.1) GO TO 208
        DO 8989 K=1,NG
8989    XLOGL(IOUNT)=XLOGL(IOUNT)+M(K)*ALOG(T(K))
208     CONTINUE
C       For each unclassified entity, compute the Mahalanobis distance
C       between it and the estimated mean for each group
        DO 205 JJ=1,NIND
        GUM=0.0
        DO 800 K=1,NG
        AL(K)=0.0
        DO 805 I=1,NATT
        DO 805 J=1,NATT
        AL(K)=AL(K)+(X(JJ,I)-YMU(K,I))*V(K,I,J)*(X(JJ,J)-YMU(K,J))
805     CONTINUE
```

```
C     Apply Huber's weight function if robust solution required
      IF (NROB.NE.1) GO TO 902
      TAU(K,JJ)=SQRT(AL(K))
      IF (TAU(K,JJ).LT.PP) WW(K,JJ)=1.0
      IF (TAU(K,JJ).GE.PP) WW(K,JJ)=PP/TAU(K,JJ)
      GO TO 903
902   WW(K,JJ)=1.0
903   CONTINUE
      IF (AL(K).GT.75.0) GO TO 8055
      AL(K)=-0.5*AL(K)
      AL(K)=EXP(AL(K))/(SQRT(DV(K))*(2.*PI)**(NATT/2.0))
      GO TO 898
8055  AL(K)=0.0
898   CONTINUE
      C=1.0E-30
      IF (T(K).LT.C.OR.AL(K).LT.C) GO TO 800
      GUM=GUM+T(K)*AL(K)
800   CONTINUE
C     Compute current estimates of posterior probabilities of group
C     membership (W)
      IF (GUM.EQ.0.0) GO TO 8110
      DO 810 K=1,NG
      IF (T(K).LT.C.OR.AL(K).LT.C) GO TO 901
      W(JJ,K)=T(K)*AL(K)/GUM
      GO TO 810
901   W(JJ,K)=0.0
810   CONTINUE
      XLOGL(IOUNT)=XLOGL(IOUNT)+ALOG(GUM)
      GO TO 205
8110  DO 8101 K=1,NG
8101  W(JJ,K)=0.0
      WRITE(22,116) IOUNT,JJ
116   FORMAT (/2X,'In loop ',I3,' the estimated mixture density is ',
     1'zero ',/4X,'for the observation on unclassified entity ',I6)
205   CONTINUE
C     Test for exit from loop
      IF (IOUNT.LE.10) GO TO 230
      LAST=IOUNT-10
      ALIM=0.0001*XLOGL(LAST)
      DIFF=XLOGL(IOUNT)-XLOGL(LAST)
      IF (ABS(DIFF).LE.ABS(ALIM)) GO TO 990
      IF (IOUNT.GT.149) GO TO 985
230   CONTINUE
C     Compute new estimate of mixing proportion (T) for each group
      DO 666 K=1,NG
      WSUM(K)=0.0
      XSUM(K)=0.0
      WSUMM(K)=0.0
      YSUM(K)=0.0
      YYSUM(K)=0.0
      DO 667 JJ=1,NIND
      WSUM(K)=WSUM(K)+W(JJ,K)*WW(K,JJ)*WW(K,JJ)
      XSUM(K)=XSUM(K)+W(JJ,K)
      WSUMM(K)=WSUMM(K)+W(JJ,K)*WW(K,JJ)
667   CONTINUE
      DO 767 JJ=1,MIND
      YSUM(K)=YSUM(K)+WY(K,JJ)
      YYSUM(K)=YYSUM(K)+WY(K,JJ)*WY(K,JJ)
767   CONTINUE
C     The computation of the estimates of the mixing proportions
C     depends on whether the classified data were obtained by mixture
C     sampling (NSAM=1) or by separate sampling (NSAM=2)
```

```
      IF (NSAM.EQ.1) GO TO 675
      T(K)=XSUM(K)/NIND
      GO TO 666
675   T(K)=(XSUM(K)+M(K))/(NIND+MIND)
666   CONTINUE
C     Compute new estimates of group means (YMU)
      DO 671 K=1,NG
      DO 671 J=1,NATT
      SUM(K,J)=0.0
      SUMY(K,J)=0.0
      DO 670 JJ=1,NIND
      SUM(K,J)=SUM(K,J)+X(JJ,J)*W(JJ,K)*WW(K,JJ)
670   CONTINUE
      DO 770 JJ=1,MIND
      SUMY(K,J)=SUMY(K,J)+Y(JJ,J)*WY(K,JJ)
770   CONTINUE
      YMU(K,J)=(SUM(K,J)+SUMY(K,J))/(WSUMM(K)+YSUM(K))
671   CONTINUE
C     Compute new estimate of covariance matrix for each group
      DO 870 K=1,NG
      DO 871 J=1,NATT
      DO 871 I=1,J
      V(K,I,J)=0.0
      VY(K,I,J)=0.0
871   CONTINUE
      DO 872 JJ=1,NIND
      DO 873 J=1,NATT
      DO 873 I=1,J
      V(K,I,J)=V(K,I,J)+(X(JJ,I)-YMU(K,I))*(X(JJ,J)-YMU(K,J))*W(JJ,K)
     1*WW(K,JJ)*WW(K,JJ)
873   CONTINUE
872   CONTINUE
      DO 772 JJ=1,MIND
      DO 773 J=1,NATT
      DO 773 I=1,J
      V(K,I,J)=V(K,I,J)+(Y(JJ,I)-YMU(K,I))*(Y(JJ,J)-YMU(K,J))
     1*WY(K,JJ)*WY(K,JJ)
773   CONTINUE
772   CONTINUE
      DO 673 J=1,NATT
      DO 673 I-1,J
      V(K,I,J)=(V(K,I,J)+VY(K,I,J))/(WSUM(K)+YYSUM(K))
      V(K,J,I)=V(K,I,J)
673   CONTINUE
870   CONTINUE
C     Obtain inverse and determinant of each estimated covariance
C     matrix
      CALL GDET(NATT,NG,V,V1,DV,IER)
      IF (IER.NE.0) RETURN
      DO 410 K=1,NG
      DO 410 J=1,NATT
      DO 410 I=1,NATT
      XVAR(K,J,I)=V(K,J,I)
410   V(K,J,I)=V1(K,J,I)
      IOUNT=IOUNT+1
      GO TO 1111
985   CONTINUE
      WRITE (22,115)
115   FORMAT (//2X,'Note: This sample did not converge in 150 ',
     1'iterations.',/8X,'However the program will continue to ',
     2'print results ',/8X,'obtained from the last cycle estimates.')
990   CONTINUE
```

```
      WRITE (22,611) IOUNT,XLOGL(IOUNT)
611   FORMAT (//2X,'In loop ',I3,' log likelihood is ',F15.3)
      WRITE (22,612)
612   FORMAT (/4X,'Estimate of mixing proportion for each group')
      WRITE (22,102) (T(K),K=1,NG)
102   FORMAT (2X,10F7.3)
C     Determine partition of unclassified entities into NG groups
      DO 30 K=1,NG
30    XCC(K)=0.0
      IOUT=0
      DO 50 I=1,NIND
      MAX=1
      DO 40 K=2,NG
40    IF (W(I,K).GT.W(I,MAX)) MAX=K
      GUM=0.0
      DO 45 K=1,NG
45    GUM=GUM+W(I,K)
      IF (GUM.EQ.0.0) GO TO 48
      XCC(MAX)=XCC(MAX)+W(I,MAX)
      IDT(I)=MAX
      GO TO 50
48    IOUT=IOUT+1
      DO 49 J=1,NATT
      Z(IOUT,J)=X(I,J)
49    IOL(IOUT)=I
50    CONTINUE
      WRITE (22,101)
101   FORMAT (//2X,'Unclassified entity: Final estimates of posterior',
     1' probabilities of group membership'/)
      DO 60 I=1,NIND
60    WRITE (22,103) I,(W(I,K),K=1,NG)
103   FORMAT (2X,I6,2X,10F7.3)
      WRITE (22,107)
107   FORMAT (//2X,'Resulting partition of the unclassified entities ',
     1'into NG groups ')
      WRITE (22,108) (IDT(I),I=1,NIND)
108   FORMAT (2X,10I4)
      DO 310 K=1,NG
310   N(K)=0
      DO 320 I=1,NIND
      K=IDT(I)
      IF (K.EQ.0) GO TO 320
      N(K)=N(K)+1
320   CONTINUE
      WRITE (22,113)
113   FORMAT (//2X,'Number assigned to each group')
      WRITE (22,114) (N(K),K=1,NG)
114   FORMAT (2X,10I6)
      IF (IOUT.EQ.0) RETURN
      WRITE (22,110) IOUT
110   FORMAT (//2X,'There were ',I4,' unclassified entities with ',
     1'observations ',/4X,'for which the estimated mixture density ',
     2'is zero.',/4X,'These data are listed with the entity number ',
     3'first.'/)
      DO 300 I=1,IOUT
300   WRITE (22,111) IOL(I),(Z(I,J),J=1,NATT)
111   FORMAT (2X,I6,2X,10F9.3)
      RETURN
      END
      PROGRAM KTMM
C     Program for fitting a mixture of univariate normal distributions
C     in order to cluster a set of treatment means.  Data on the
```

```
C     treatments are available from an experimental design with error
C     variance, sigma**2; that is, the variance of a treatment sample
C     mean is sigma**2 divided by the number of replications of the
C     treatment.  The necessary data are the treatment sample means
C     and an independent estimate of sigma**2 with IDF degrees of
C     freedom.
C
C     The input file contains the following:
C     NIND                 (number of means to be clustered)
C     (XBAR(I),I=1,NIND)   (treatment sample means)
C     (NR(I),I=1,NIND)     (replication number for each mean)
C     VARS IDF             (error variance and its degrees of freedom)
C     NG                   (number of groups)
C     (IDT(K),K=1,NG)      (initial grouping of the means)
C        This allows free field input but for ease of reading use
C        ten numbers per line for I=1,NIND.
C     NOTE:  The initial partition of the treatment means requires at
C            least one mean to be assigned to each group.
C
C     The output file contains the results of fitting the above model
C     starting from an initially specified partition of the means.
C     It includes the final probabilistic clustering and the
C     consequent partition of the NIND treatments into NG groups.
C
      COMMON NIND,NG,XBAR(60),XMU(10),NR(60),VARS,IDF,VARE,
     1T(10),IDT(10),W(60,10)
      CALL GIN
      CALL MWO
      CALL LOOP
      STOP
      END


      SUBROUTINE GIN
C     This subroutine reads the data, and calculates estimates of the
C     mean and mixing proportion for each group corresponding to the
C     initial partition of the NIND means into NG groups.
      COMMON NIND,NG,XBAR(60),XMU(10),NR(60),VARS,IDF,VARE,
     1T(10),IDT(10),W(60,10)
      DIMENSION N(10)
      READ (21,*) NIND
      READ (21,*) (XBAR(I),I=1,NIND)
      READ (21,*) (NR(I),I=1,NIND)
      READ (21,*) VARS, IDF
      WRITE (22,105) VARS, IDF
105   FORMAT (2X,'Initial estimate of error variance ',E15.8,
     1' with ',I6,' d.f.')
      VARE=VARS
      READ (21,*) NG
      READ (21,*) (IDT(I),I=1,NIND)
      WRITE (22,109)
109   FORMAT (//2X,'Initial partition as specified by input')
      WRITE (22,110) (IDT(I),I=1,NIND)
110   FORMAT (2X,10I4)
      DO 16 K=1,NG
      N(K)=0
16    XMU(K)=0.0
C     Compute estimates of mean and mixing proportion for each group
      DO 50 JJ=1,NIND
      K=IDT(JJ)
      N(K)=N(K)+1
50    XMU(K)=XMU(K)+XBAR(JJ)
```

```
      DO 35 K=1,NG
      XMU(K)=XMU(K)/N(K)
      T(K)=N(K)
35    T(K)=T(K)/NIND
      WRITE (22,106)
106   FORMAT (//2X,'Estimate of each group mean')
      WRITE (22,103) (XMU(K),K=1,NG)
103   FORMAT (2X,10F12.5)
      WRITE (22,108)
108   FORMAT(//2X,'Proportion from each group as specified by input')
      WRITE (22,107) (T(K),K=1,NG)
107   FORMAT (5X,10F7.3)
      RETURN
      END


      SUBROUTINE MWO
C     This subroutine calculates, for each treatment, the estimated
C     posterior probabilities of group membership.
      COMMON NIND,NG,XBAR(60),XMU(10),NR(60),VARS,IDF,VARE,
     1T(10),IDT(10),W(60,10)
      DIMENSION DIFF(60,10),D(60,10),DIV(60)
      DO 40 I=1,NIND
      DO 10 K=1,NG
10    DIFF(I,K)=XBAR(I)-XMU(K)
      DO 20 K=1,NG
20    D(I,K)=T(K)*EXP(-NR(I)*DIFF(I,K)**2/(2.0*VARE))
      DIV(I)=0.0
      DO 25 K=1,NG
25    DIV(I)=DIV(I)+D(I,K)
      DO 30 K=1,NG
30    W(I,K)=D(I,K)/DIV(I)
40    CONTINUE
      RETURN
      END


      SUBROUTINE LOOP
C     This subroutine uses the EM algorithm from a specified starting
C     value to find a solution of the likelihood equation.
      COMMON NIND,NG,XBAR(60),XMU(10),NR(60),VARS,IDF,VARE,
     1T(10),IDT(10),W(60,10)
      DIMENSION WSUM(10),SUM(10),XLOGL(100),XCC(10),WWSUM(10)
      PI=3.141592653
      IOUNT=1
1111  CONTINUE
C     Compute the log likelihood
      A1=NIND/2.0*ALOG(2.0*PI)
      A2=NIND/2.0*ALOG(VARE)
      A3=IDF*(ALOG(VARE)+VARS/VARE)/2.0
      A4=0.0
      C1=0.0
      DO 630 I=1,NIND
      XNR=NR(I)
      C1=C1+0.5*ALOG(XNR)
      B2=0.0
      DO 635 K=1,NG
      B3=(XBAR(I)-XMU(K))**2*NR(I)/(2.0*VARE)
635   B2=B2+T(K)*EXP(-B3)
630   A4=A4+ALOG(B2)
      Z=IDF/2.0
      C2=(Z-1.0)*ALOG(VARS)
```

```
      C3=GAMMAZ(Z)
      XLOGL(IOUNT)=A4-A2-A1-A3+C1+C2-C3
C     Test for exit from loop
      IF (IOUNT.LE.10) GO TO 230
      LAST=IOUNT-10
      ALIM=0.0005*XLOGL(LAST)
      DIFF=XLOGL(IOUNT)-XLOGL(LAST)
      IF (ABS(DIFF).LE.ABS(ALIM)) GO TO 990
      IF (IOUNT.GT.74) GO TO 985
230   CONTINUE
C     Compute new estimate of mixing proportion (T) for each group
C     using the estimates of the posterior probabilities (W) from
C     previous loop
      DO 666 K=1,NG
      WSUM(K)=0.0
      WWSUM(K)=0.0
      DO 667 JJ=1,NIND
      WWSUM(K)=WWSUM(K)+W(JJ,K)*NR(JJ)
667   WSUM(K)=WSUM(K)+W(JJ,K)
666   T(K)=WSUM(K)/NIND
C     Compute new estimates of group means (XMU)
      DO 668 K=1,NG
      SUM(K)=0.0
      DO 670 JJ=1,NIND
670   SUM(K)=SUM(K)+XBAR(JJ)*W(JJ,K)*NR(JJ)
668   XMU(K)=SUM(K)/WWSUM(K)
C     Compute new estimate of error variance (VARE)
      A5=0.0
      DO 930 K=1,NG
      DO 930 I=1,NIND
930   A5=A5+W(I,K)*(XBAR(I)-XMU(K))**2*NR(I)
      VARE=(IDF*VARS+A5)/(IDF+NIND)
C     Compute current estimates of posterior probabilities of group
C     membership (W)
      CALL MWO
      IOUNT=IOUNT+1
      GO TO 1111
985   CONTINUE
      WRITE (22,115)
115   FORMAT (//2X,'This sample did not converge in 75 iterations')
990   CONTINUE
      WRITE (22,611) IOUNT, XLOGL(IOUNT)
611   FORMAT (//2X, 'In loop ',I3,' log likelihood is ',F15.3)
      WRITE (22,612)
612   FORMAT (/4X,'Estimate of mixing proportion for each group')
      WRITE (22,333) (T(K),K=1,NG)
333   FORMAT (5X,10F7.3)
      WRITE (22,101)
101   FORMAT (//2X,'Treatment: Final estimates of posterior ',
     1'probabilities of group membership'/)
      DO 60 I=1,NIND
60    WRITE (22,102) I,(W(I,K),K=1,NG)
102   FORMAT (2X,I6,2X,10F7.3)
C     Determine partition of treatments into NG groups
      DO 30 K=1,NG
30    XCC(K)=0.0
      DO 50 I=1,NIND
      MAX=1
      DO 40 K=2,NG
40    IF (W(I,K).GT.W(I,MAX)) MAX=K
      XCC(MAX)=XCC(MAX)+W(I,MAX)
50    IDT(I)=MAX
```

```
      WRITE (22,107)
107   FORMAT (//2X,'Resulting partition of the treatment means ',
     1'into NG groups ')
      WRITE (22,108) (IDT(I),I=1,NIND)
108   FORMAT (2X,10I4)
      WRITE (22,106)
106   FORMAT (//2X,'Estimate of each group mean')
      WRITE (22,113) (XMU(K),K=1,NG)
113   FORMAT (2X,10(F7.3,2X))
      WRITE (22,117) VARE
117   FORMAT (//2X,'Estimate of error variance ',E15.8)
      RETURN
      END


      FUNCTION GAMMAZ(S)
C     This subroutine computes the natural log of the complete
C     gamma function G(p).
      P=S
      IF (S.EQ.0.5) GO TO 6
      IF (P.EQ.1.0) GO TO 4
      NQ=P
      IF (P.EQ.NQ) GO TO 1
C     Test if N is integer; 1=yes
      SUM=0.0
2     P=P-1.0
      SUM=SUM+ALOG(P)
      IF (P.GT.1.0) GO TO 2
      SUM=SUM+0.572364943
      GO TO 3
1     SUM=0.0
5     P=P-1.0
      SUM=SUM+ALOG(P)
      IF (P.GT.1.5) GO TO 5
3     GAMMAZ=SUM
      RETURN
6     GAMMAZ=0.572364943
      RETURN
4     GAMMAZ=0.0
      RETURN
      END


      PROGRAM K3MM
C     Program for fitting a mixture of normal distributions with
C     arbitrary covariance matrices to three-mode three-way data.
C     The original program, which was similar in form to the others
C     in the appendix, was converted to single dimension arrays by
C     Ian DeLacy from the Department of Agriculture, University of
C     Queensland.  The use of such arrays is more economical and
C     allows more flexibility in applying the program to problems
C     of different sizes.
C     The first mode is taken to be the class of entities to be
C     clustered.  The second mode is a class of some characteristics
C     of the entities, referred to as the attributes.  The
C     third mode consists of the different levels of some factor,
C     referred to as the environments.  For each entitiy, each
C     attribute is measured in each environment, yielding a three-mode
C     three-way array representation of the data.
C
C     To enable it to initially set up the storage arrays, the
C     program interactively asks for the following:
```

```
C         NIND  (The number of entities to be clustered)
C         NATT  (The number of attributes)
C         NBL   (The number of environments)
C         NG    (The number of groups)
C         NITM  (The maximum number of iterations for the EM algorithm)
C      The input file contains the following:
C      (X(I,J,L),J=1,NBL)
C      ------------------   (NIND rows of data, each having NBL values
C      ------------------    per line.  This is repeated for each of
C      ------------------    NATT attributes.)
C      ------------------
C      (IDT(I),I=1,NIND)    (initial grouping of the entities)
C         This allows free field input but for ease of reading use
C         ten numbers per line for I=1,NIND.
C      NOTE:  The initial partition of the entities requires at
C             least NATT+1 entities to be assigned to each group.
C
C      The output file contains the results of fitting the above model
C      starting from an initially specified partition of the data.
C      It includes the final probabilistic clustering and the
C      consequent partition of the NIND entities into NG groups.
C
       DIMENSION Z(10000)
       LIMZ=10000
       CALL DIMMIX(Z,LIMZ)
       STOP
       END


       SUBROUTINE DIMMIX(Z,LIMZ)
C      This subroutine sets up the storage arrays.
       DIMENSION Z(*)
       WRITE (6,1000)
1000   FORMAT (2X,'Type in number of entities'/,
      18X,'number of attributes'/,
      28X,'number of environments'/,
      38X,'number of groups'/,
      48X,'maximum number of iterations (usually 100)')
C      If the environments are considered as blocks then the three-way
C      genotype by attribute by environment data can be viewed as
C      multiattribute RCBD data
       READ (5,*) NT,NA,NE,NG,NI
       NTEA=NT*NE*NA
       NTA=NT*NA
       NEA=NE*NA
       NGA=NG*NA
       NGAA=NG*NA*NA
       NGAE=NG*NA*NE
       NAA=NA*NA
       NGG=NG*NG
       NTG=NT*NG
       NM=MAX0(NT,NE,NA,NG)
       N0=1
       N1=N0+NTEA
       N2=N1+NTA
       N3=N2+NEA
       N4=N3+NG
       N5=N4+NGA
       N6=N5+NGAE
       N7=N6+NGAA
       N8=N7+NG
       N9=N8+NGAA
```

```
      N10=N9+NGAE
      N11=N10+NT
      N12=N11+NG
      N13=N12+NA
      N14=N13+NGAE
      N15=N14+NAA
      N16=N15+NA
      N17=N16+NG
      N18=N17+NTG
      N19=N18+NG
      N20=N19+NGA
      N21=N20+NGAA
      N22=N21+NI
      N23=N22+NG
      N24=N23+NGAE
      N25=N24+NM
      IF (N25.GT.LIMZ) GO TO 999
      CALL MIX(Z(N0),Z(N1),Z(N2),Z(N3),Z(N4),Z(N5),Z(N6),Z(N7),Z(N8),
     1Z(N9),Z(N10),Z(N11),Z(N12),Z(N13),Z(N14),Z(N15),Z(N16),Z(N17),
     2Z(N18),Z(N19),Z(N20),Z(N21),Z(N22),Z(N23),Z(N24),
     3NT,NE,NA,NG,NI)
      RETURN
999   WRITE (6,1001) N25,LIMZ
1001  FORMAT (2X,'Dimension of Z array too small'/
     12X,'It should be increased to ',I6,' from ',I6)


      RETURN
      END


      SUBROUTINE MIX(X,TRMEAN,BLMEAN,T,XMU,GAMMA,
     1XVAR,DV,V,DEV,IDT,N,XMEAN,ASUM,E,IP,AL,W,WSUM,SUM,
     2V1,XLOGL,XCC,EXTA,TPE,NIND,NBL,NATT,NG,NITM)
C     This subroutine calls the other subroutines as necessary and then
C     writes out some summary results.
      DIMENSION X(*),TRMEAN(*),BLMEAN(*),T(*),XMU(*),GAMMA(*)
      DIMENSION XVAR(*),DV(*),V(*),DEV(*),IDT(*),N(*),XMEAN(*),ASUM(*),
     1E(*),IP(*),AL(*),W(*),WSUM(*),SUM(*),
     2V1(*),XLOGL(*),XCC(*),EXTA(*),TPE(*)
      CALL GIN(XVAR,IDT,N,XMEAN,ASUM,TPE,NIND,NG,NATT,NBL,X,
     1TRMEAN,BLMEAN,T,XMU,GAMMA,V,DV,IER,E,IP)
      IF (IER.NE.0) GO TO 605
      CALL LOOP(V,DV,AL,W,WSUM,SUM,V1,IDT,XLOGL,XCC,EXTA,E,IP,TPE,
     1IER,NITM,NIND,NG,NATT,NBL,X,TRMEAN,T,XMU,XVAR,GAMMA)
      IF (IER.NE.0) GO TO 605
      WRITE (22,106)
106   FORMAT (//2X,'Estimated mean (as a row vector) for each group')
      DO 20 K=1,NG
      M1=A2(K,1,NATT)
      M2=A2(K,NATT,NATT)
20    WRITE (22,103) (XMU(M),M=M1,M2)
103   FORMAT (2X,10F13.6)
      DO 30 K=1,NG
      WRITE (22,105) K
105   FORMAT (//2X,'Estimated covariance matrix for group ',I12)
      DO 30 J=1,NATT
      DO 31 I=1,J
      M=A3(K,I,J,NATT,NATT)
31    TPE(I)=XVAR(M)
30    WRITE (22,103) (TPE(I),I=1,J)
      WRITE (22,110)
```

```
110    FORMAT (//2X,'Estimated group by environment effect for ',
      1'each attribute'/)
       DO 50 L=1,NATT
       DO 51 I=1,NG
       DO 52 K=1,NBL
       M3=A3(I,K,L,NBL,NATT)
52     TPE(K)=GAMMA(M3)
51     WRITE (22,103) (TPE(K),K=1,NBL)
50     WRITE (22,115)
115    FORMAT (2X)
       DO 60 I=1,NG
       DO 60 K=1,NBL
       DO 60 L=1,NATT
       M1=A3(I,K,L,NBL,NATT)
       M2=A2(I,L,NATT)
60     DEV(M1)=GAMMA(M1)-XMU(M2)
       WRITE (22,111)
111    FORMAT (//2X,'Deviation from estimated group mean of ',/,
      14X,'estimated group by environment effect for each attribute '/)
       DO 70 L=1,NATT
       DO 71 I=1,NG
       DO 72 K=1,NBL
       M=A3(I,K,L,NBL,NATT)
72     TPE(K)=DEV(M)
71     WRITE (22,103) (TPE(K),K=1,NBL)
70     WRITE (22,115)
       GO TO 620
605    WRITE (22,101) IER
101    FORMAT (//2X,'Terminal error in MATINV as IERR =',I6)
620    RETURN
       END


       SUBROUTINE GIN(XVAR,IDT,N,XMEAN,ASUM,TPE,NIND,NG,NATT,NBL,X,
      1TRMEAN,BLMEAN,T,XMU,GAMMA,V,DV,IER,E,IP)
C      This subroutine reads the data and calculates estimates of the
C      mean, covariances matrix, and mixing proportion for each group,
C      corresponding to the initial partition of the entities into NG
C      groups.
       DIMENSION X(*),TRMEAN(*),BLMEAN(*),XMU(*),T(*),GAMMA(*),
      1V(*),DV(*),E(*),IP(*)
       DIMENSION XVAR(*),IDT(*),N(*),XMEAN(*),ASUM(*),TPE(*)
       DO 10 L=1,NATT
       DO 10 J=1,NIND
       READ (21,*) (TPE(K),K=1,NBL)
       DO 10 K=1,NBL
       M=A3(J,K,L,NBL,NATT)
10     X(M)=TPE(K)
       READ (21,*) (IDT(I),I=1,NIND)
       WRITE (22,109)
109    FORMAT (//2X,'Initial partition as specified by input')
       WRITE (22,110) (IDT(I),I=1,NIND)
110    FORMAT (2X,10I4)
C      Compute estimate of mean for each group
71     DO 15 K=1,NG
       N(K)=0
       DO 15 J=1,NATT
       M=A2(K,J,NATT)
       XMU(M)=0.0
       DO 15 I=1,NATT
       M=A3(K,I,J,NATT,NATT)
15     XVAR(M)=0.0
```

```
      DO 26 L=1,NATT
      XMEAN(L)=0.0
      DO 25 J=1,NIND
      M=A2(J,L,NATT)
25    TRMEAN(M)=0.0
      DO 26 K=1,NBL
      M=A2(K,L,NATT)
26    BLMEAN(M)=0.0
      DO 24 I=1,NG
      DO 24 K=1,NBL
      DO 24 L=1,NATT
      M=A3(I,K,L,NBL,NATT)
24    ASUM(M)=0.0
      DO 28 J=1,NIND
      DO 28 L=1,NATT
      DO 27 K=1,NBL
      M1=A2(J,L,NATT)
      M2=A3(J,K,L,NBL,NATT)
27    TRMEAN(M1)=TRMEAN(M1)+X(M2)
28    TRMEAN(M1)=TRMEAN(M1)/REAL(NBL)
      DO 32 K=1,NBL
      DO 32 L=1,NATT
      DO 31 J=1,NIND
      M1=A2(K,L,NATT)
      M2=A3(J,K,L,NBL,NATT)
31    BLMEAN(M1)=BLMEAN(M1)+X(M2)
32    BLMEAN(M1)=BLMEAN(M1)/REAL(NIND)
      DO 36 L=1,NATT
      DO 33 K=1,NBL
      M=A2(K,L,NATT)
33    XMEAN(L)=XMEAN(L)+BLMEAN(M)
      XMEAN(L)=XMEAN(L)/REAL(NBL)
36    CONTINUE
      WRITE (22,106)
106   FORMAT (//2X,'Estimated mean (as a row vector) for each group')
      DO 20 JJ=1,NIND
      K=IDT(JJ)
      N(K)=N(K)+1
      DO 20 J=1,NATT
      DO 19 KB=1,NBL
      M1=A3(K,KB,J,NBL,NATT)
      M2=A3(JJ,KB,J,NBL,NATT)
      M3=A2(K,J,NATT)
      M4=A2(JJ,J,NATT)
19    ASUM(M1)=ASUM(M1)+X(M2)
20    XMU(M3)=XMU(M3)+TRMEAN(M4)
C     Compute estimate of mixing proportion for each group
      DO 35 K=1,NG
      DO 30 J=1,NATT
      M=A2(K,J,NATT)
30    XMU(M)=XMU(M)/FLOAT(N(K))
      T(K)=N(K)
      T(K)=T(K)/FLOAT(NIND)
      DO 34 J=1,NATT
      M1=A2(K,J,NATT)
34    TPE(J)=XMU(M1)
35    WRITE (22,103) (TPE(J),J=1,NATT)
103   FORMAT (2X,10F13.6)
C     Compute estimate of group by environment effect for each
C     attribute
      DO 80 I=1,NG
      DO 80 K=1,NBL
```

```
      DO 80 L=1,NATT
      M=A3(I,K,L,NBL,NATT)
80    GAMMA(M)=ASUM(M)/FLOAT(N(I))
C     Compute estimate of covariance matrix for each group
      DO 202 J=1,NIND
      I=IDT(J)
      DO 202 L=1,NATT
      DO 202 LD=1,L
      DO 202 K=1,NBL
      M1=A3(I,L,LD,NATT,NATT)
      M2=A3(J,K,L,NBL,NATT)
      M3=A3(I,K,L,NBL,NATT)
      M4=A3(J,K,LD,NBL,NATT)
      M5=A3(I,K,LD,NBL,NATT)
202   XVAR(M1)=XVAR(M1)+(X(M2)-GAMMA(M3))*(X(M4)-GAMMA(M5))
      DO 201 I=1,NG
      DO 201 L=1,NATT
      DO 201 LD=1,L
      M1=A3(I,L,LD,NATT,NATT)
      M2=A3(I,LD,L,NATT,NATT)
      XVAR(M1)=XVAR(M1)/FLOAT(NBL*N(I))
201   XVAR(M2)=XVAR(M1)
      DO 60 K=1,NG
      WRITE (22,105) K
105   FORMAT (//2X,'Estimated covariance matrix for group ',I4)
      DO 60 J=1,NATT
      DO 61 I=1,J
      M=A3(K,I,J,NATT,NATT)
61    TPE(I)=XVAR(M)
60    WRITE (22,103) (TPE(I),I=1,J)
C     Obtain inverse and determinant of each estimated covariance
C     matrix
      CALL GDET(XVAR,V,DV,E,IP,IER,NG,NATT)
      WRITE (22,108)
108   FORMAT (//2X,'Proportion from each group as specified by input')
      WRITE (22,107) (T(K),K=1,NG)
107   FORMAT (5X,10F7.3)
      WRITE (22,120)
120   FORMAT (/2X,'Estimated group by environment effect for each ',
     1'attribute')
      DO 82 L=1,NATT
      DO 81 I=1,NG
      DO 83 K=1,NBL
      M=A3(I,K,L,NBL,NATT)
83    TPE(K)=GAMMA(M)
81    WRITE (22,103) (TPE(K),K=1,NBL)
82    WRITE (22,114)
114   FORMAT (2X)
      RETURN
      END


      SUBROUTINE GDET(XVAR,V,DV,E,IP,IER,NG,NATT)
C     This subroutine reads all covariance matrices, then calls
C     MATINV, which inverts a matrix and calculates its determinant,
C     for each covariance matrix in turn.
      DIMENSION XVAR(*),V(*),DV(*),E(*),IP(*)
      DO 40 K=1,NG
      DO 20 J=1,NATT
      DO 20 I=1,J
      M1=A3(K,I,J,NATT,NATT)
      M2=A2(I,J,NATT)
```

```
      M3=A2(J,I,NATT)
      E(M2)=XVAR(M1)
20    E(M3)=E(M2)
      TOL=0.000001
      CALL MATINV(E,NATT,NATT,NATT,DET,IP,TOL,IER)
      IF (IER.NE.0) RETURN
      DO 35 J=1,NATT
      DO 35 I=1,J
      M1=A3(K,I,J,NATT,NATT)
      M2=A3(K,J,I,NATT,NATT)
      M3=A2(I,J,NATT)
      V(M1)=E(M3)
      V(M2)=V(M1)
35    CONTINUE
      DV(K)=DET
40    CONTINUE
      END


      SUBROUTINE LOOP(V,DV,AL,W,WSUM,SUM,V1,IDT,XLOGL,XCC,EXTA,E,IP,
     1TPE,IER,NITM,NIND,NG,NATT,NBL,X,TRMEAN,T,XMU,XVAR,GAMMA)
C     This subroutine uses the EM algorithm from a specified starting
C     value to find a solution of the likelihood equation.
      DIMENSION X(*),TRMEAN(*),T(*),XMU(*),XVAR(*),GAMMA(*)
      DIMENSION V(*),DV(*),AL(*),W(*),WSUM(*),SUM(*),V1(*),IDT(*),
     1XLOGL(*),XCC(*),TPE(*),EXTA(*),E(*),IP(*)
      PI=3.141592653
      IOUNT=1
      XT=FLOAT(NBL)/2.0
      XTT=XT*NATT*ALOG(2.0*PI)
1111  CONTINUE
C     Compute the log likelihood
      XLOGL(IOUNT)=0.0
      DO 205 J=1,NIND
      GUM=0.0
      DO 800 I=1,NG
      AL(I)=0.0
      DO 805 L=1,NATT
      DO 805 LD=1,NATT
      DO 805 K=1,NBL
      M1=A3(J,K,L,NBL,NATT)
      M2=A3(I,K,L,NBL,NATT)
      M3=A3(I,L,LD,NATT,NATT)
      M4=A3(J,K,LD,NBL,NATT)
      M5=A3(I,K,LD,NBL,NATT)
      AL(I)=AL(I)+(X(M1)-GAMMA(M2))*V(M3)*(X(M4)-GAMMA(M5))
805   CONTINUE
      ALL=-0.5*AL(I)-XT*ALOG(DV(I))-XTT
      IF (ALL.LT.-80.) ALL=-80.0
      AL(I)=EXP(ALL)
      GUM=GUM+T(I)*AL(I)
800   CONTINUE
      XLOGL(IOUNT)=XLOGL(IOUNT)+ALOG(GUM)
C     Compute current estimates of posterior probabilities of group
C     membership (W)
      DO 810 I=1,NG
      M1=A2(J,I,NG)
810   W(M1)=T(I)*AL(I)/GUM
205   CONTINUE
C     Test for exit from loop
      IF (IOUNT.LE.10) GO TO 230
      LAST=IOUNT-10
```

```
      ALIM=0.0001*XLOGL(LAST)
      DIFF=XLOGL(IOUNT)-XLOGL(LAST)
      IF (ABS(DIFF).LE.ABS(ALIM)) GO TO 990
      IF (IOUNT.GT.(NITM-1)) GO TO 985
230   CONTINUE
C     Compute new estimate of mixing proportion (T) for each group
      DO 666 K=1,NG
      WSUM(K)=0.0
      DO 667 JJ=1,NIND
      M=A2(JJ,K,NG)
667   WSUM(K)=WSUM(K)+W(M)
666   T(K)=WSUM(K)/FLOAT(NIND)
C     Compute new estimates of group means (XMU)
      DO 671 K=1,NG
      DO 671 J=1,NATT
      M1=A2(K,J,NATT)
      SUM(M1)=0.0
      DO 670 JJ=1,NIND
      M2=A2(JJ,J,NATT)
      M3=A2(JJ,K,NG)
670   SUM(M1)=SUM(M1)+TRMEAN(M2)*W(M3)
671   XMU(M1)=SUM(M1)/WSUM(K)
C     Compute new estimates of group by environment effects
C     for each attribute
      DO 530 K=1,NBL
      DO 530 L=1,NATT
      DO 530 I=1,NG
      M1=A3(I,K,L,NBL,NATT)
      EXTA(M1)=0.0
      DO 525 J=1,NIND
      M2=A2(J,I,NG)
      M3=A3(J,K,L,NBL,NATT)
525   EXTA(M1)=EXTA(M1)+W(M2)*X(M3)
530   GAMMA(M1)=EXTA(M1)/WSUM(I)
C     Compute new estimate of covariance matrix for each group
      DO 870 I=1,NG
      DO 873 L=1,NATT
      DO 873 LD=1,L
      M1=A3(I,L,LD,NATT,NATT)
      M2=A3(I,LD,L,NATT,NATT)
      V(M1)=0.0
      DO 872 K=1,NBL
      DO 872 J=1,NIND
      M3=A3(J,K,L,NBL,NATT)
      M4=A3(I,K,L,NBL,NATT)
      M5=A2(J,I,NG)
      M6=A3(J,K,LD,NBL,NATT)
      M7=A3(I,K,LD,NBL,NATT)
872   V(M1)=V(M1)+(X(M3)-GAMMA(M4))*W(M5)*(X(M6)-GAMMA(M7))
      V(M1)=V(M1)/(NBL*WSUM(I))
873   V(M2)=V(M1)
870   CONTINUE
C     Obtain inverse and determinant of each estimated covariance
C     matrix
      CALL GDET(V,V1,DV,E,IP,IER,NG,NATT)
      IF (IER.NE.0) RETURN
      DO 410 K=1,NG
      DO 410 J=1,NATT
      DO 410 I=1,NATT
      M=A3(K,J,I,NATT,NATT)
      XVAR(M)=V(M)
410   V(M)=V1(M)
```

```
551    IOUNT=IOUNT+1
       GO TO 1111
985    CONTINUE
       WRITE (22,115) NITM
115    FORMAT (//2X,'Note: This sample did not converge in ',I6,
      1'iterations.'/8X,'However the program will continue to print ',
      2'results ',/8X,'obtained from the last cycle estimates.')
990    CONTINUE
       WRITE (22,611) IOUNT, XLOGL(IOUNT)
611    FORMAT (//2X, 'In loop ',I3,' log likelihood is ',F15.3)
       WRITE (22,612)
612    FORMAT (/4X,'Estimate of mixing proportion for each group')
       WRITE (22,333) (T(K),K=1,NG)
333    FORMAT (5X,10F7.3)
       WRITE (22,101)
101    FORMAT (//2X,'Entity: Final estimates of posterior ',
      1'probabilities of group membership'/)
       DO 60 I=1,NIND
       M1=A2(I,1,NG)
       M2=A2(I,NG,NG)
60     WRITE (22,102) I,(W(M),M=M1,M2)
102    FORMAT (2X,I6,2X,10F7.3)
C      Determine partition of entities into NG groups
       DO 30 K=1,NG
30     XCC(K)=0.0
       DO 50 I=1,NIND
       MAX=1
       DO 40 K=2,NG
       M1=A2(I,K,NG)
       M2=A2(I,MAX,NG)
40     IF (W(M1).GT.W(M2)) MAX=K
       XCC(MAX)=XCC(MAX)+W(M2)
50     IDT(I)=MAX
       WRITE (22,107)
107    FORMAT (//2X,'Resulting partition of the entities into NG groups')
       WRITE (22,108) (IDT(I),I=1,NIND)
108    FORMAT (2X,10I4)
860    RETURN
       END


       SUBROUTINE MATINV(E,NR,NC,NMAX,DET,IP,TOL,IERR)
       DIMENSION E(*),IP(*)
C      This is a matrix inversion subroutine originally written by
C      I. Oliver of the University of Queensland Computer Centre in
C      September 1966 in FORTRAN IV.  It was revised by J. Williams in
C      April 1967; then revised and updated by I. Oliver in November
C      1969.  Ref. UQ D4.505
C      In February 1986, it was converted to use one dimensional arrays
C      by Ian DeLacy of the Department of Agriculture, University of
C      Queensland.
C
C      Initialization
       IERR=0
       IF (NR.LE.NMAX.AND.NR.GT.0.AND.NC.GT.0) GO TO 5
       IERR=2
       RETURN
5      DET=1.0
       DO 10 I=1,NR
10     IP(I)=I
       DO 170 K=1,NR
C      Search for pivot
```

```
      IR=K
      AMAX=0.0
      DO 20 I=K,NR
      M=A2(I,K,NR)
      IF (ABS(E(M)).LE.ABS(AMAX)) GO TO 20
      AMAX=E(M)
      IR=I
20    CONTINUE
      DET=DET*AMAX
C     Record pivot row
      I=IP(K)
      IP(K)=IP(IR)
      IP(IR)=I
C     Test for zero pivot
      IF (AMAX.NE.0.) GO TO 70
      IERR=1
      RETURN
C     Move selected pivot row to pivot row
70    IF (IR.EQ.K) GO TO 100
      DO 90 J=1,NC
      M1=A2(IR,J,NC)
      M2=A2(K,J,NC)
      TEMP=E(M1)
      E(M1)=E(M2)
90    E(M2)=TEMP
      DET=-DET
C     Transform matrix
100   DO 110 J=1,NC
      M=A2(K,J,NC)
110   E(M)=E(M)/AMAX
      DO 150 I=1,NR
      IF (I.EQ.K) GO TO 150
      DO 140 J=1,NC
      IF (J.EQ.K) GO TO 140
      M1=A2(I,J,NC)
      M2=A2(I,K,NR)
      M3=A2(K,J,NC)
      T1=E(M1)
      T2=E(M2)*E(M3)
      T3=T1-T2
      IF (ABS(T3).LT.(ABS(T1)+ABS(T2))*TOL)T3=0.0
      E(M1)=T3
140   CONTINUE
150   CONTINUE
      DO 160 I=1,NR
      M1=A2(I,K,NR)
      M2=A2(K,K,NR)
160   E(M1)=-E(M1)/AMAX
      E(M2)=1.0/AMAX
170   CONTINUE
C     Realign inverse
      DO 220 K=1,NR
      DO 180 J=1,NR
      IF (K.EQ.IP(J)) GO TO 190
180   CONTINUE
190   IF (J.EQ.K) GO TO 220
      DO 210 I=1,NR
      M1=A2(I,J,NR)
      M2=A2(I,K,NR)
      TEMP=E(M1)
      E(M1)=E(M2)
210   E(M2)=TEMP
```

```
      I=IP(K)
      IP(K)=IP(J)
      IP(J)=I
220   CONTINUE
      IF (IERR.EQ.1) IERR=3
      IF (IERR.EQ.2) IERR=0
      RETURN
      END


      FUNCTION A2(I,J,NC)
      A2=(I-1)*NC+J
      RETURN
      END


      FUNCTION A3(I,J,K,NJ,NK)
      IT=(I-1)*NJ+J
      A3=(IT-1)*NK+K
      RETURN
      END
```

# Author Index

*Italic numbers give the page on which the complete reference is listed.*

# Subject Index